T0164113

GREAT NATIONAL CRAFTSMEN

www.royalcollins.com

GREAT NATIONAL CRAFTSMEN

By Production Team of Great National Craftsmen

Books Beyond Boundaries

ROYAL COLLINS

Great National Craftsmen

By Production Team of *Great National Craftsmen*
Edited by Kong Defang
Translated by Liu Xianshu

First published in 2021 by Royal Collins Publishing Group Inc.
Groupe Publication Royal Collins Inc.
BKM Royalcollins Publishers Private Limited

Headquarters: 550-555 boul. René-Lévesque O Montréal (Québec) H2Z1B1 Canada
India office: 805 Hemkunt House, 8th Floor, Rajendra Place, New Delhi 110008

Original Edition © New World Press

ISBN: 978-1-4878-0847-1 (hardcover)
ISBN: 978-1-4878-0848-8 (paperback)

To find out more about our publications, please visit www.royalcollins.com

Dedicated
to
All the Hardworking People!

Model workers are elites of the nation and role models for the people. The great craftsmen are highly proficient representatives of the workforce. The All-China Federation of Trade Unions shall, in joint efforts with other organizations, build a platform for model workers and craftsmen to better play their role and provide the soil for the fostering of more such workers and craftsmen.

—President Xi Jinping at a meeting with the new leadership of the All-China Federation of Trade Unions, October 29, 2018

This year, China has been taking on a new look through joint efforts to promote "Made in China," "Made by China" and "Created in China." We successfully carried out the Chang'e-4 lunar probe; our second aircraft carrier set sail on its maiden voyage; our domestically-made large amphibious aircraft performed its first water launch; and the BeiDou Satellite Navigation System went global. I would like to extend my respects to each and every scientist, engineer, craftsman and all the other participants in the building of our country!

—President Xi Jinping's New Year Speech for 2019, December 31, 2018

A Brief Introduction to the Series *Great National Craftsmen*

Great National Craftsmen, a key exclusive program of News Center of CCTV, was broadcast on *CCTV News* and News Channel CCTV in 2015. As of May 2019, the program has been aired for seven seasons.

The program tells touching stories of the front-line representatives of technicians from different industries. These dedicated workers focus on their careers with great perseverance, holding the country's development and the nation's rejuvenation above personal benefits. Being content with a simple life, they devote all efforts to their careers. Instead of pursing personal gains with their expertise, they work for the prosperity of the nation. They strive to pass down their expertise and cultivate more successors. Although many of them have not received higher education and have only got vocational training, they have become masters in their fields through hard work and practice. Just as Xi Jinping has said, they are "workers with wisdom and skills capable of invention and creation, and they contribute to realizing the Chinese Dream with their hard work."

Great National Craftsmen extols the spirit of craftsmanship. Following the broadcasts, the program has evoked strong empathy and received popular recognition. It has received high affirmation from the central leadership. In 2015, the team which produced the program won the National Labor Medal issued by the All-China Federation of Trade Unions. In 2016, *Great National Craftsmen* received the first prize of the 26th China News Award.

PREFACE

Great Work from Great Efforts[1]

By Xu Qiang, deputy director of News Center, CCTV

On International Labor Day, CCTV broadcast its key program titled *Great National Craftsmen*. After its debut on the Morning News channel, the eight-episode program appeared in different columns of the news channel. *CCTV News* allotted a total of 39 minutes out of its compact content arrangement to broadcast a shortened version of the 8 episodes, paying tribute to the great craftsmen of China.

The planning of *Great National Craftsmen* started on April 14, 2015, when Nie Chenxi, president of CCTV, presided over research at the CCTV Center, and consulted the staff. He put forward the requirements of enhancing the quality of the news output of CCTV. International Labor Day was not a hotspot occasion for CCTV, when routine reports on domestic and international news, holiday and outing events and model workers of China would generally suffice. However, the leaders of CCTV required that we should do more and do it better. Sun Yusheng, vice-president of CCTV, proposed that news reports on Labor Day should be treated as a special task. He pointed out that in-depth research and design should be conducted to make *Great National Craftsmen* unique.

---

1. On May 27, 2015, CCTV held a symposium on *Great National Craftsmen*. The participants included leaders from the Publicity Department of the CPC Central Committee, All-China Federation of Trade Unions, State Administration of Press, Publications, Radio, Film and Television and CCTV, eight representatives of great craftsmen, the team for the program, experts and media representatives. The speeches by two members of the production team of *Great National Craftsmen* at the symposium are selected as the preface of this book.

I. Advocating craftsmanship: A proper theme of far-reaching significance

The planning for Labor Day was in itself a big theme, and the depth of thinking of the planners was directly related to the impact of the program. The CCTV News Channel played a vital guiding role in advocating the core values of socialism. Centering on the theme of labor, we decided to focus on the labor values which should not be, but were, neglected. Nowadays, many people no longer hold the belief that "every trade has its champion." Instead of striving to become more capable in work, many people lack the patience to improve themselves, and are overly concerned with material wealth. In many industries, there is a lack of technical talent with a down-to-earth attitude and dedication to perfection in their professions. Our era calls for a return to the plain labor spirit and less pompousness.

In China's education system, the main obstacle to the development of vocational education is that people's opinion of technical talents needs to be improved. By focusing on excellent models, we can make more people realize that vocational education can create top-level talents who can be regarded as a nation's treasure. After an in-depth analysis, the theme of the program became clearer: guiding public opinion on craftsmanship, making more people respect craftsmanship and pursue expertise in their fields, and advocating a social trend of respecting diligent and creative work.

After we decided upon the theme of *Great National Craftsmen,* we had to decide how to carry it out with positive feedback. First, we had to decide currently which industries in China deserved the "Great National" status. Second, we had to come up with the criteria of "Craftsmen." In our opinion, the program *Great National Craftsmen* should focus on the two opposite concepts of "big" and "small," i.e., big industries with inspiring achievements and "small" figures with touching stories. We believe that throughout the course of its civilization of several thousand years, China did not lack the inheritance and advocacy of craftsmanship; rather, what we lacked was discovery, i.e., discovery of the top craftsmanship in contemporary manufacturing industries, revealing the high quality of "Made-in-China," exploring the craftsmanship spirit of modern workers and telling their touching stories. Our reporters and editors, who have long been keeping up with related fields, quickly found their subjects. By exploring eight leading industries, we discovered the top-talented technicians working in their respective fields. Apart from their talent and expertise, they have come up with innovations, contributing greatly to the development of their fields. What is more, they all have profound affection for the Chinese nation.

II. Improving the operation mechanism: Exercising strict control to producing quality products

The key to ensuring program quality was to establish an effective organizational structure and operational mechanism.

Great National Craftsmen adopted a general director responsibility system, which means the general director had the right to deploy human resources throughout the early interviews, post-stage production, new media, program promotion, etc. He was responsible for supervising and controlling the entire progress of the program, arranging the broadcasting, following the feedback, and so on.

As a flagship program of the Labor Day reports, *Great National Craftsmen,* instead of the conventional work mode based on individual departments and teams, drew leading staff from different departments of the News Center into the core production team of the program. Eight groups of senior reporters were selected from the Economic News Department. Professional editing staff from News Channel's editing department came to offer their expertise. The team from the program *Sincere Communication* of the Visual Arts Department participated in the topic design and provided guidance in the shooting. There were art editors responsible for the packaging design, and personnel from the New Media News Department were responsible for designing feature products suitable for new media communication. With the establishment of the project system, the staff could focus completely on the program and take more initiatives.

In order to ensure the quality of each episode, a time table and task diagram were set up at the very beginning of the production of the program. The production process was divided into many links. The chief director conducted task evaluation and quality check at key time junctures, realizing a whole-process control throughout the topic design, interview and follow-up, post-stage editing, program review and promotion.

III. Intensive production and broadcasting: Early input and wide coverage

Apart from shooting the program, we also had to study the market and audience. As well as the conventional TV mass communication, efforts were also needed in the precise and socialized communication of particular users of new media. In order to realize the optimal communication effect, the rigid teaching input mode had to be replaced by a more appealing and infiltrating mode which offered the audience more choices. In terms of the making and broadcasting of the program and its promotion, Huang Chuanfang, deputy chief editor and managing director of the News Center made two requirements: "being early" and "wide coverage."

"Being early" meant good timing of the release time. We decided to seize the opportunity to achieve a preemptive effect before the public and other media entered the "Labor Day model." *Great National Craftsmen* was originally scheduled to be released one week prior to Labor Day. After the one-week introduction, the program would reach a climax on May 1, May 2 and May 3. However, as a 8.1-magnitude earthquake struck Nepal on April 25 and affected China's Tibet Autonomous Region, we analyzed the situation and postponed the broadcast time of the program until April 29, with a further release on the first three days of the Labor Day holiday. Results showed that the release prior to May 1 enhanced the popularity of the program.

Also, "Being early" meant early promotion. The promotion program for *Great National Craftsmen* was released one week in advance. The slogan "*Great National Craftsmen*: Building Their Dreams with Dedication" well delivered the theme of the program. The promotion program was of such a high standard and good quality that it aroused strong interest and expectations among the audience.

"Being wide" meant the dimension and depth of release. Not general, but well-targeted. Firstly, in terms of content, among the theme reports for Labor Day that year, the flagship program *Great National Craftsmen* was released in advance and for several days continuously. In contrast, other programs, such as the more sentimental programs of *Workers' Poems, Reflection on Working, New Entrepreneurs, Maker as I Am* targeted at young people, adopted a menu-type design targeting specific audience groups. Secondly, in terms of release, on the basis of audience analysis and public opinion studies, we enhanced the release frequency of *Great National Craftsmen* on *CCTV News, Morning News, News Alive* and other programs.

Besides, different versions were made considering the requirements of different programs, which enhanced the pertinence and effective reach of each. Thirdly, "being wide" meant wide coverage on multi-media platforms. After the release on TV, *Great National Craftsmen* was released on multiple media, with related content and features covering the Weibo, WeChat accounts and the APP of CCTV News simultaneously.

Great National Craftsmen was a project with a tight schedule and demanding work load. Throughout the process of identifying the shooting subjects to finalizing the shooting, many editors and reporters didn't have adequate sleep for days. I want to express thanks to them for their hard work of high efficiency, which has made *Great National Craftsmen* released as scheduled with a great success.

How *Great National Craftsmen* Came to Be Made

By Yue Qun, News Center of CCTV

Great National Craftsmen, which won the first prize of the China News Award, was a special program of the CCTV News Center, released on *CCTV News* during the May Day and National Day holidays in 2015. The program tells the stories of 17 top technicians in different fields, and extols workers who "hold up the Chinese Dream with hard work." It reflects the inheritance and promotion of the traditional craftsman spirit, advocates hard, honest and creative work, and shows the high-quality image of "Made-in-China."

Frankly speaking, we had not expected that *Great National Craftsmen* would become such a big hit. I remember that when the first season of the program was broadcast, we entrusted a third party big-data consultancy firm to carry out a feedback survey via the Internet. To our surprise, the program got a 91% positive feedback, which even surpassed that of *A Bite of China 2* on Douban. We wondered why *Great National Craftsmen* could have won such a high score. I would like to talk about our opinions in terms of the conception and operation of the program:

I. Looking for the conjunction of the program and the era

Our initial inspiration for *Great National Craftsmen* came from a conversation with an entrepreneur who told us a story of an old man making a transistor radio manually. He said that he was deeply touched by the old man's spirit of pursuing perfection.

We were also deeply touched. When it comes to craftsmanship spirit, people will most often mention Swiss watchmakers. Although China is the No.1 manufacturer in the world, Chinese tourists rush to buy rice cookers and toilet seat covers in other countries. Why? Could it be that there lacks the craftsmanship spirit in Chinese workers, or have we been ignoring its existence because the society is more concerned with pursuing material gains?

In mid-April 2015, the News Center of CCTV launched the program planning for the May Day holiday in advance. Our idea of seeking craftsmanship got approval from the department as well as from the News Center. The planning work of the News Center, which was carried out well ahead, left us enough time to think about the orientation of the program. The title evolved from Great Craftsmen of the Nation, Great Craftsmen and finally to Great National Craftsmen. It was not merely a matter of different combinations of Chinese characters, but reflected the consensus after several rounds of discussion. Our consensus was to highlight the respect for workers represented by craftsmen, and advocate their spirit.

We spent nearly two thirds of our limited time on the program orientation and character selection. When the major projects at the top of China's manufacturing industry were combined with basic-level craftsmen with superb skills, a good conjunction point between theme report and appeal of the era was found, which laid a solid foundation for the success of the program.

II. Producing inspiring TV programs

It was not easy to tell convincingly the stories in *Great National Craftsmen*, as the interviewees had one thing in common: They are all taciturn. In one episode, an apprentice commented that his master was "exceptional at work" and "exceptionally taciturn outside work."

In fact, the craftsmen in our program have a rich spiritual world, and most of their inner world is reflected in their work. Despite the tight schedule of shooting, our reporters managed to conduct in-depth communication with them to discover the plainest yet most touching details and aspects in their lives.

At first, our subjects were not used to the company of our reporters, who spent almost all the time with them except for sleeping and going to bathroom. While the former was working, the reporters would be there, observing their expressions, movements, words and even tones of voice. When they took a rest, the reporters would be there, too, having meals with them at the same table and chatting with them. Gradually, the skilled workers opened their hearts to the reporters. One even admitted that it was the first time that he shed tears before someone he was not familiar with.

The in-depth interviews revealed the admirable qualities of the craftsmen like polishing up the shining facets of a diamond. Their words were so touching as to reach the bottom of the audience's hearts. "As long as one learns the words of any of them by heart, one will benefit throughout life," one of the viewers said.

Another challenge in telling the stories of the craftsmen well lay in the camera work. As the masters all specialize in quite demanding and meticulous work – for example, one of them deals with a precision of one tenth of the diameter of a hair – it was quite challenging to show the details to the audience. What's more, their operations are mostly quite monotonous, such as welding, grinding, drilling, etc. They engage in such operations for hours, and our cameramen kept shooting for hours. Most of the time, the cameramen and reporters were trying to find the best angle and studying the light, so there were actually not many scenes. In order to capture a subject's hand movements, one cameraman tied a camera to the subject's wrist. Another used a 100-macro-lens camera.

Gao Fenglin, who specializes in welding the "heart" of rockets, told his interviewer that the light from the argon arc welding he used was five times more harmful than that of electric welding. During the interview, he kept reminding the reporter not to look at the light, as it was quite harmful to the eyes. However, in order to record Gao's superb welding skills, the reporter insisted on shooting without a protective shield. When the shooting ended, the reporter had to close his eyes for a while for some rest and adjustment. Thanks to the dedication and hard work of the reporters, the stories of the great masters were told in such a vivid and touching way.

III. Marching ahead with the craftsmanship spirit

The shooting of *Great National Craftsmen* was an unforgettable spiritual journey for our reporters. One reporter wrote: "I had been searching for the essence of the subject's spirit while we were making the program, and I got to know its value and greatness in the process. At the same time, I felt a great driving force to surpass myself to produce a quality program under the pressing schedule and in an exhausted state."

Why the pressing schedule? Because two thirds of the time were spent in determining the theme and selecting the interview subjects. As a result, the reporters had only a few days for each interview, including their commuting, contacting the interviewee, the interview, shooting and editing. It was almost "mission impossible." One of our colleagues commented that to produce such a high-quality program in such a short period of time, there must have been some spiritual drive.

Indeed, in the process of shooting, the reporters also kept learning the craftsmanship spirit. Many worked almost around the clock. They communicated with one another in the WeChat work group of the program from 5 a.m. to 1 a.m. or 2 a.m. of the next day. There was brainstorming over various issues, big and small. For example, interview requirements, shooting, and adjustment of a shot or a line. There were also heated discussions on many details. Everyone was striving to produce a perfect program, just like the dedicated craftsmen.

In order to make an excellent program, everyone worked with the utmost perseverance. One of the reporters, who had a heart problem, always had some special pills in her pocket in case she felt uncomfortable in the middle of an interview. Another reporter's wife had just undergone an operation, but he had no time to take care of her; he asked his mother-in-law for help while he himself hurried to his interview. In the LNG ship, as high as ten storeys, a female reporter, who was thin and weak, climbed up and down in steel shoes weighing several kilos every day until the skin on her heels was worn off. Some young female staff who were responsible for the processing and polishing of the program worked 24 hours for several days on end.

Through producing *Great National Craftsmen*, the reporters had a chance to appraise themselves. They had better ideas as to how to be more rigorous, efficient and meticulous in their future work, as well as how to hold fast to their commitment to journalism. Not long ago, as the representative of the winners of the China News Award, I had the opportunity of hearing General Secretary Xi Jinping's speech on that occasion in person. I felt all the more that as news professionals we should exercise strict self-discipline, take a down-to-earth approach, and strive to make perfect products with the same steadfast commitment as the great craftsmen. Only in this way can we produce good news and get the affirmation of the audience.

China has abundant work force, yet it is not a technical workforce power. It is a big industrial nation, but it has a long way to go to become a global industrial power. It is the demand of the era for our manufacturing industry to go through a transition, upgrade and innovate, for which ordinary workers and front-line workers are indispensable. Through *Great National Craftsmen,* we strive to call for more concern from society about the value of craftsmen as well as to form favorable public opinion for the elevation of the social status of craftsmen. By elevating craftsmen's value in terms of cognitive concept and evaluation criteria, the craftsmen's image will be better established, and more substantive thinking on the establishment of an evaluation and benefit system for craftsmen will be stimulated.

We will carry on making programs like *Great National Craftsmen,* to evoke, through our lasting efforts, more social respect for professional skills. We hope that the skills of the masters can be passed down so that the craftsmanship spirit will light up more people's dreams.

CONTENTS

GAO FENGLIN

Rocket "Heart" Welder

Profile

Gao Fenglin is an aerospace special fusion welding technician of Capital Aerospace Machinery Company, China Aerospace Science and Technology Corporation. He is the leader of the Gao Fenglin Team of the company as well as a national model worker. Nearly 40% of the "hearts" of China's rockets, including the Long March 5, were welded under his supervision. He has tackled more than 200 difficult technical problems. He has made multiple breakthroughs in many large key projects, for example, new-type materials production, new technology, new structure and new methods, especially in the research and development of the new high-thrust engine. He solves difficult problems with great courage, extraordinary insight, rigorous reasoning and highly honed skills. He creatively applies his understanding of the welding process to automation, intelligent control and other soft-type processes. Gao Fenglin has made outstanding contributions to the modernization of national defense, and aerospace science and technology.

Gao Fenglin, welder of Capital Aerospace Machinery Company, China Aerospace Science and Technology Corporation

Preview

The carrier rocket is the prerequisite for human beings' voyages in and exploration of space, and the engine is the heart of the carrier rocket. Gao Fenglin, a special fusion welding technician of Capital Aerospace Machinery Company, China Aerospace Science and Technology Corporation, has been a welder for over 30 years. He best represents the meaning of "craftsman" with his superb skills and pursuit of perfection. He also embodies the responsibility and mission of an aerospace worker with his own persistence.

O n December 2, 2013, the Chang'e-3 lunar detector was launched from the Xichang Satellite Launch Center. It was launched on a voyage to the moon on the Long March 3B, China's flagship rocket. The engine or "heart" of the Long March 3B was welded by Gao Fenglin and his colleagues.

The distance from the earth to the moon is 380,000 km. The width of a soldered joint on a rocket engine is 0.16 mm. The time error allowed for welding is 0.1 s. Welding might seem simple, but the welding of aerospace hardware requires rigorous judgement in positioning, welding angle and strength. It is a comprehensive test in terms of brain, eyesight, physical strength and willpower of the welders.

Gao Fenglin, 53 years old, is the leading welding technician of rocket engines of the China Aerospace Science and Technology Corporation. His latest challenge is welding the engine for the Long March 5, China's new-generation large carrier rocket now under development.

The engine of the Long March 5 has the largest thrust of any engine developed by China so far. It is fueled by liquid hydrogen and oxygen. It is vital for ensuring the carrying capacity of the Long March 5. The temperature of the combustion is as high as above 3,000°C. Inside the engine's nozzle are several hundred hollow

Gao Fenglin welds a rocket engine.

pipes with a diameter of several millimeters and a pipe wall as thin as 0.33 mm each. The welding seams of the pipes, although as thin as hairs, have a total length that equals the perimeter of a standard football field. In order to fix these tiny pipes, Gao Fenglin needs to weld them a total of more than 30,000 times, with great accuracy.

Gao Fenglin must ensure that his each and every operation step is correct and perfect, and he has only one chance. When the engine is working, propellant of nearly –200°C flows into the tiny tubes to cool it down. Should there be any defect in any welding joint, the flames in the engine will tear the nozzle apart, and the entire rocket will explode.

"We must gaze without blinking, especially at the the tiny welding joints," Gao Fenglin said. "We must gaze with the utmost concentration. I can keep my eyes wide open without a single blink if I have to."

"How can you not blink for ten minutes?" I asked.

"It's no big deal. I can show you."

Gao Fenglin smiled confidently. His confidence came from long-term diligent study and hard practice. No mistake is allowed in aerospace manufacturing, and workers must start by acquiring solid basic skills. A small flaw in the welding of the engine, which powers the rocket and thus is called the "heart" of the rocket, might be disastrous. Therefore, the welding requires not only superb skills, but also a meticulous attitude and strict self-discipline. Gao Fenglin learned this from his senior colleagues on his first day at work.

Chen Jifeng, Gao Fenglin's mentor, was one of the pioneers of special welding after 1949. Special welding, as a core technology of national pillar industries, is generally used in welding special materials that are hard in nature and prone to damage from the external environment. To master this sophisticated welding technique, one shall have not only superb skills, but also professional dedication. When Gao Fenglin was an apprentice, Chen Jifeng started by teaching him the basic gestures. For instance, one shall not rest one's hand on the workbench while operating. Rather, one should lift one's arm and keep it suspended like practicing calligraphy. In this way, the elbow and wrist can move with great agility to ensure that the welding torch can reach the most difficult welding area. Besides, to ensure the welding quality, one must keep one's hand extremely steady. Another risk to welding stability is the welder's breathing. The slight movement of body caused by breathing may affect the welding quality.

"How can one practice breathing?" The journalist asked Chen Jifeng, who was almost 90 years old.

"Just hold it!"

From gesture to breathing, the rigorous training was beyond Gao Fenglin's expectation. There is one thing that Gao Fenglin will never forget:

"One day, I welded a test piece. The rule was to check it from both sides after welding. When I took up the test piece, I felt it a little bit hot, so I threw it on floor immediately."

Gao Fenglin did not expect to be severely criticized by his master for such a small thing.

"You should respect your work," his master said to him seriously.

To respect the work means having a correct work attitude. On hearing these words, Gao Fenglin had a new understanding of his work. He practiced hard. Even when he was queuing up for meals in the canteen, he practiced with chopsticks.

When Gao Fenglin puts on his welding mask, he enters a completely different world. From behind the mask, he fixes his eyes on the workpiece in the flickering

Gao Fenglin shares his experience with colleagues.

5

electric arc. The solid welding wire melts into a soft fluid. He is totally focused, completely immersed in his own small world, as he has been doing for years.

For Gao Fenglin, the emergence of new rocket models means new technical challenges and breakthroughs. Once he stayed in the workshop day and night and barely slept for a whole month.

In the first few days, some colleagues worked with him until mid-night. Several days later, only one or two stayed until three o'clock. A few days later, Gao Fenglin was the only one left in the workshop.

At night, only Gao Fenglin was working in the workshop of several hundred square meters. All that could be heard was the humming of the welding torch, and all that could be seen was the faint light of welding. He was totally focused on the tiny dots of molten metal, ignoring everything else.

"You stayed up late for so many days. Weren't you tired?" I asked him.

"Of course I was, but I set my mind on carrying on. I had to succeed no matter how hard it was. You see, no one in my family suffers from baldness, but I do."

Gao Fenglin is always hard-working like this. He has solved one technological problem in aerospace manufacturing welding after another, and become the top welding expert in the industry. Once an X-ray photo inspection showed that there might be internal welding defects in an aerospace welding part. Gao Fenglin, from his years of experience, judged that the welding was reliable and the part was of good quality. A further test confirmed Gao Fenglin's judgment. Then people began to realize that Gao Fenglin's eyes were better than X-rays.

In the interview, Gao Fenglin kept emphasizing that this story was somewhat exaggerated and that the secret of his correct judgment was the basic welding skills and accumulated experience. Gao said that his achievement was due to spending 80% of his total time on working and 15% on studying.

So only 5% of his time remains unoccupied. "I leave that time for my family," Gao Fenglin said. Although he is very busy and works overtime frequently, he spends time with his family as much as possible. But there are not many opportunities.

Gao Fenglin has a lovely daughter, but he doesn't have much time to spend with her. Sometimes he takes her to school or brings her home when school is over. Gao Fenglin recalled that when he first went to the kindergarten to pick his daughter up, she ran to him happily. She was so excited that the kindergarten teachers were surprised. When they saw Gao Fenglin, they immediately understood.

"No wonder the girl is happy; you are such a rare guest!"

Many enterprises valued Gao Fenglin's superb skills and tried to hire him, offering high salaries. Some even offered a salary several times higher and two apartments in Beijing. Gao Fenglin admitted that the terms were attractive, and his wife would have liked him to go, too.

Unlike China today launching more than 20 rockets a year, it was indeed the winter of high-end manufacturing decades ago. At that time, the aerospace and military manufacturing sectors were at a low ebb, and enterprises didn't have many orders. Although Gao Fenglin was already a renowned master in the industry, he still earned a meager salary. Many skilled workers left, some of whom even abandoned their skills. However, Gao Fenglin chose to stay in his nearly-deserted workshop.

"I am so proud when I see the engines that we have built power the rockets to send satellites into space. This pride cannot be measured by money. If I left, I probably could not get this sense of fulfillment."

It is this sense of fulfillment that made Gao Fenglin stick to his post. Over 35 years, more than 130 Long March rockets boosted by engines he welded have successfully flown into space, accounting for more than half of the total Long

Gao Fenglin checks welding quality.

7

March rockets launched in China. These include the Long March 5, Long March 6 and Long March 7, a new generation of carrier rockets that are becoming an important symbol of China's march to become a space power.

Today, although a lot of advanced technologies are applied in spacecraft manufacturing, welding is still one of the most important core technologies. In developing rockets, efforts contributed by academicians, professors and senior engineers are indispensable. However, the transition from the blueprint to material object relies on numerous welding points and the dedication of many ordinary workers.

Every day, Gao Fenglin is the last one to leave work. Before he leaves, he always looks around the workshop. The shiny rocket engines are standing quietly in the workshop. These are the fruits of his and his colleagues' hard work, and they call them affectionately "golden babies." They place great hope on the engines.

"Look, it is shiny and like a perfect art piece. It is beautiful! It is our golden baby. We brought it into the world."

In Gao Fenglin's opinion, one should constantly follow the development course of all things, and pursue perfection.

Craftsman Spirit Is "Competing" with Oneself

Gao Fenglin is the first hero in the *Great National Craftsmen* series.

Before I interviewed Gao Fenglin, the program *Great National Craftsmen* was only a plan on paper. I had only two days to produce a sample. I knew that the first crab-eater was brave, but on the other hand, there are always uncertainties in the first attempt.

Despite the uncertainties, my colleagues and I started working. We used cameras, but not video cameras. At that time, many people thought that shooting a program to be broadcast on CCTV with ordinary cameras was somehow amateurish. However, this practice almost became normal for the *Great National Craftsmen* series.

It was not an easy job to make an attractive program on Gao Fenglin. First of all, Gao Fenglin is calm and not very emotional. His work involves sophisticated technology. But I was not used to elevating the concept of his job to the utmost human level at first.

Secondly, Gao Fenglin's work was monotonous. We filmed almost everything on the first day. It was then that we understood what kind of subjects we were dealing with in shooting *Great National Craftsmen*. Our protagonist had only two tools: a welding mask and a welding torch. His operations seem simple, i.e., first observing the workpiece closely and thoroughly, and then welding. To outsiders, although Gao Fenglin had the honor and aura of great master, his work was actually dull and repetitive. How could we discover the craftsmen's creativity, and how could we represent it?

I was a little bit frustrated.

I discussed with my colleague He Cheng if we could try to shoot the program from the perspective of behind the mask. I found that shooting programs was similar to Edison's inventing the light bulb. There were thousands of kinds of filament materials. One needed not only inspiration but also constant attempts to find the most applicable material.

Thus we designed the scene of the green halo of welding flashing against the tail flame of a flying rocket at the beginning of the episode.

When a rocket is launched, the combustion time of its tail flame is about 1,700 seconds. In order to ensure stable combustion in those 1,700 seconds, Gao Fenglin has to weld 30,000 times, or stare at the welding halo 30,000 times over a period of about one month.

Welding with such high intensity for one month is strenuous both mentally and physically. When Gao Fenglin puts aside his welding torch and takes off his mask, he still stares at the workpiece, murmuring as if he is still thinking about something.

"Did you ever fail?" I asked him.

"I am not allowed to fail!"

But what if he is confronted with constant failures throughout a month?

I recalled how Gao Fenglin asked his colleagues to leave when they were challenging a technological breakthrough. When he mentioned that more and more colleagues left, there wasn't any helplessness or disappointment in his expression or voice. He did not make any complaint. He knew he must succeed, and believed that he would surely succeed. He faced not only physical challenges, but also challenges to his willpower.

I began to understand why Gao Fenglin always mentioned "human beings" in interviews.

If humans meet aliens in space one day, the latter may scan the engines of our rockets to judge the level of our civilization, and the welding seams of Gao Fenglin might be the benchmark for defining human civilization.

If intelligent machines replace human beings in the industrial field one day, the first thing they "learn" with their intelligent chips might be to feel and copy, with their cold mechanical arms, how Gao Fenglin welds rocket engines.

Lastly, more perseverance and concentration lead one closer to perfection, which is exactly the strongest internal motivation for human beings to create a brilliant civilization.

Maybe this is what craftsmen mean to our world.

By WU JIE, reporter of
China Media Group & CCTV News Center

MENG JIANFENG

A Life of Carving

Profile

Meng Jianfeng is the senior technician of Beijing Orafi Jewelry Co., Ltd., Beijing Gongmei Group. With more than 20 years' experience, he has received many honorary titles and awards, including National Model Worker in professional ethics construction, "Model of state-owned enterprises · Beijing model" and Capital Labor Medal. Meng Jianfeng's story is reported in the CCTV feature program *Great National Craftsmen*. His works include the medals for the scientists involved with China's atomic and hydrogen bombs, and man-made satellites, the "Shenzhou" series aerospace hero medals, APEC Summit gifts, gifts for the "Belt and Road Forum for International Cooperation." Meng Jianfeng has been making unremitting efforts to improve his skills. He is an excellent model in the industry dedicated to carrying forward traditional culture and inheriting traditional craftsmanship.

Meng Jianfeng, senior technician of Beijing Orafi Jewelry Co., Ltd., Beijing Gongmei Group

Preview

Chisel engraving is a Chinese craft with a history of nearly 3,000 years in which artisans chisel out various relief patterns on gold, silver, copper or other metals. During the APEC Summit in Beijing in 2014, an artware of chisel engraving named *Hemei* ("Harmony & Happiness") in the form of a silk scarf held in a tray was offered as a national gift to the wives of the heads of state participating the summit. The art piece was made by the chisel engraving artisan Meng Jianfeng.

Chisel engraving is a traditional craft of China with a history of nearly 3,000 years. When the APEC Summit was held in Beijing in November 2014, the heads of state attending the Summit were surprised by the ancient Chinese chisel engraving technique. Following social customs, Chinese President Xi Jinping and his wife Peng Liyuan prepared national gifts representing the history, culture and craftsmanship of China for the couples attending the summit. Among the gifts there was a chisel engraving artwork in the form of a silver silk scarf held in a golden tray. The artwork was named *Hemei*, which meant harmony and happiness in Chinese. On seeing the artwork, the guests couldn't help touching the silk scarf, but no one could pick it up. When they realized that the silk scarf and tray were one piece, they all marveled at the exquisite artwork, and expressed profuse praise. This exquisite artwork was made by Meng Jianfeng of China Gongmei Group.

Meng Jianfeng lives in Beijing. Every morning, he hurries to work like the other commuters of the city.

Meng Jianfeng's company is located near the Beijing Olympic Park. Every day he goes through half of the city to work. The company separates him from the hustle and bustle outside temporarily. When he enters the workshop where he has worked for more than 20 years, his mind becomes tranquil immediately.

His workplace is a two-story factory building built in the 1980s. It lacks a modern factory atmosphere. It is of a grey color and has a simple layout, and the air is filled with the faint smell of metals being worked. It is in this old factory that Meng Jianfeng and his colleagues have made various exquisite artworks, including national gifts for the APEC, medals for experts involved in atomic and hydrogen bombs, and China's first man-made satellite and the "Shenzhou" aerospace series, as well as other traditional art pieces.

On the first floor there are small rooms along a long corridor. Inside the rooms, or workshops to be exact, the technicians are busy smelting, filigreeing, profiling, engraving, etc. Exquisite artworks are produced here. At present, Meng Jianfeng and his apprentices are smelting metal in a traditional way, which is the first step in making national gifts.

Meng Jianfeng is wearing an apron and thick gloves that reach his elbows. He switches on a furnace to smelt a silver ingot. Red flames are reflected on his face. Soon he sweats a lot, but he has no time to wipe the sweat off. He is totally immersed in his work.

The national gift Hemei

Meng Jianfeng works in his studio.

Meng Jianfeng's studio is on the second floor. On the door there is a plate reading "Studio of Meng Jianfeng, Senior National Artisan." The most important and most difficult steps in making national gifts are taken here.

The tools used for chisel engraving are chisels of different sizes and shapes – circle, half-moon, fine line, etc. – to produce various patterns on metal surfaces.

Meng Jianfeng holds a chisel firmly onto a silver plate with his left hand and taps it with a small hammer held in his right hand. His studio is filled with the pleasant tinkling sound, and patterns gradually appear on the silver plate. Good tools are the prerequisite for refined artwork. In order to produce beautiful patterns, the making of chisels is vital. In this trade, the jargon for chisel making is "*kaizanzi.*" In chisel engraving, various chisels are used to create different patterns. Besides, artworks of new design require innovation in chisel making, too.

The national APEC meeting gift *Hemei* features a golden-color tray and a silver silk scarf on it. The tray has a rough texture that resembles rattan weaving. The scarf, with a natural and exquisite flower pattern, seems quite soft and vivid, and is pleasant to the eye. When light shines on the artware from different angles, one can perceive the woven texture of the tray and exquisite texture of the scarf. The soft, delicate scarf and the tray of primitive simplicity form a sharp yet harmonious contrast.

Meng Jianfeng made great efforts to create the rough weaving texture of the tray and the smooth texture of the silk scarf. In particular, in order to create the refined flower pattern on the scarf, he made numerous attempts before he got the perfect effect. He racked his brain day and night to solve this problem. Even in his dreams, he thought hard about how to design chisels that could create the effect he desired.

This task lasted more than one month. In this period, Meng Jianfeng made nearly 30 chisels. It took him five days to make the smallest one under a magnifying glass. He carved more than 20 fine lines on the tip of the chisel, i.e., in an area less than 1 mm². The lines, being about 0.07 mm or approximately the diameter of a hair, were evenly distributed, completely parallel and carved to the same depth. While he was making the chisels, the fine lines could easily be erased out of the slightest carelessness, when he had to start over again.

Meng Jianfeng's chisels are kept in a pot on his desk. In order to keep the chisel tips intact, the dozens of chisels are placed with the tips pointing up. Meng Jianfeng cherishes these chisels so much that he won't allow the slightest damage to them.

Making chisels was the first step in making the national gift *Hemei*. Next was striking, which was far more challenging. Meng had to strike the material several million times, yet he could not make a mistake in even a single strike, for that would ruin all his efforts.

After hammering the silver strip into a thickness of 0.6 mm, Meng Jianfeng carved the desired patterns, bit by bit, on the silver sheet with different chisels. The tray with the weaving feature, because of its rough texture, was comparatively easy. However, the "silk scarf" on it was a big challenge.

The silk scarf was decorated with traditional flowery and scroll patterns, which symbolize wealth, auspiciousness and good fortune in Chinese culture. The scarf delivers the meaning of "harmony and happiness" and symbolizes the common wish of all economies to develop the global economy through joint efforts. In order to create these exquisite and beautiful patterns, Meng had to be extremely focused and careful.

Meng held his breath and kept his eyes wide open. He held a chisel with one hand and hit it with a small hammer held in the other. In order to fully represent

Meng Jianfeng keeps some of his specially designed chisels in a small box.

the patterns, the soft texture and casual folding effect of the silk scarf, he carved interwove fine lines and dots, even in folds less than 1 mm of the scarf.

In this process, Meng Jianfeng determined to finish each line of the pattern at one stretch without reworking, for he thought that reworking would leave traces of overlapping, and the pattern would be less smooth or natural. Although ordinary people would by no means discover that, Meng Jianfeng would not tolerate any imperfection. He said that he pursues the utmost perfection in his work. In order to achieve this perfection, each of his strokes must be steady, firm and accurate. However, as the silver sheet was only 0.6 mm thick, he had to be extremely careful not to make a hole in it. He had to remain highly focused throughout several million strokes; if he missed in one, he would start all over again.

The pursuit of perfection is the rule Meng Jianfeng has set for himself. The silk scarf and the tray had to be perfect, and the four Chinese knots under the tray had to be flawless, too. To realize this goal, Meng Jianfeng worked so hard that his right hand was covered with thick calluses.

Generally, Chinese knots are made with soft materials, for example, woven with red cords. The four Chinese knots supporting the tray in the artwork *Hemei* were

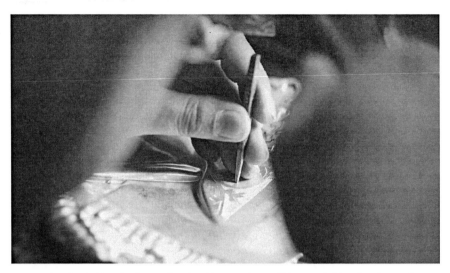

Meng Jianfeng holds a chisel in one hand and hit it with a small hammer held in the other. He keeps hitting the silver sheet with the chisel gently bit by bit in an extremely meticulous way.

supposed to be woven with silver wire. The silver became soft when heated, but it would soon become stiff again. Therefore, weaving Chinese knots with silver wire was a demanding and almost impossible job. The technicians had planned to make the Chinese knots by means of machine casting, and weld them onto the tray. However, there were tiny holes in the machine-made work, and this was not good enough for Meng Jianfeng. In the course of his devotion to craftsmanship for over 20 years, he has developed a deep affection for Chinese artwork. In his opinion, every piece of artwork has a unique life, and only 100% handmade and flawless artwork deserves the title of national gift. Therefore, he was determined to make *Hemei* a perfect piece of work which could fully and vividly represent China's 5,000-year cultural heritage.

Driven by determination, Meng Jianfeng persisted in weaving the silver wires manually. Heating, bending, then heating and bending again, interweaving, reheating, and shaping... To weave one curve of a knot, he repeated these steps dozens of times. He kept working. He worked so hard that his hands were covered with blisters. After he went home, he would puncture the blisters, cut off the dried

21

skin with a nail clipper the next day, and go on working with the knots. It took days for him to finish them. When Meng's wife recalled this later, she could not help but shed tears.

Sure enough, not only the leaders and their wives attending the APEC Summit, but also common people all marvel at the artwork *Hemei* as well as the ancient and exquisite craftsmanship of China. A sample of *Hemei* is displayed at Wangfujing Arts and Crafts Building, Beijing. Whoever sees it cannot but marvel at it. Behind the admiration is years of unremitting pursuit and dedication of its creator.

In the company, being Meng Jianfeng's apprentice is an honor, but it is also not an easy achievement. When choosing apprentices, Meng Jianfeng is concerned particularly with whether they are fond of and dedicated to craftsmanship. In his opinion, otherwise the candidate will not be totally dedicated to this work. This applies to Meng Jianfeng himself, too.

Meng Jianfeng first teaches his apprentices how to use the chisels. It may seem easy if one isn't familiar with Master Meng, who is particularly demanding. Meng Jianfeng teaches them how to hold the chisels firmly and how to use the strength of each hand. It is so exhausting that after the demonstration, his face is covered in sweat.

Meng Jianfeng is quite strict with his apprentices, which originated in his own experience. When he was new in the factory, his master taught him the same way, i.e., practicing the basic skills several hours per day. After some time, he was bored with the monotonous work, and did not want to continue. What was the use of all the repetition? He thought he was ready to learn some advanced skills.

Just as Meng Jianfeng grew disappointed, his mother, who had passed on her perseverance to him, encouraged him to hang on. She told him that since he had chosen this profession, he should stick to it, and not give up halfway. One who did everything by halves and backed off whenever there were difficulties would be good for nothing.

Meng felt the deep love from his mother, and took her advice. Every day, no matter how late he got home from work, his mother always waited for him. She never went to sleep before her son came back.

With the great confidence and courage received from his mother, Meng followed his master's instructions in practicing the basic skills every day. For one filing operation alone, he practiced for a whole year.

Although Meng Jianfeng is a senior national arts and crafts master today, he is never satisfied with himself, and is always pursuing perfection. He believes that the arts and crafts trade require painting skills, so he will go out with his wife for sketching whenever he has time. Although he was dexterous with chiseling, he was very clumsy with pencils and paintbrushes at the beginning. However, he is such a tenacious man that he will never give up. He learns painting from colleagues who do the designing. With his perseverance and diligence, he will surely create excellent paintings one day. Meng Jianfeng keeps on surpassing himself in his pursuit of perfection with his deep love and reverence for traditional Chinese craftsmanship.

"Polishing" Program with Dedication

In late April 2015, I was given a special assignment to produce an episode of the May Day special program *Great National Craftsmen*. As a macro-economic reporter, I had just finished the feature program *The Folk Track of the 1st Quarter Economy*, and was also working on another program called *The Road and Belt Initiative: Striving for Prosperity in Joint Efforts*. I had stayed up late for nearly two weeks, and was totally exhausted. I was no longer young, so I just desired a good sleep.

However, I immediately called my interviewee Meng Jianfeng, partially driven by my professional ethics and partially attracted by the program. At that time, the word craftsman was not as popular as it is now, but the phrase "Great National Craftsmen" seemed to have a mysterious appeal to me.

Thus I engaged in a new "battle." It was indeed a battle, which I had to win, not fail. I had only six days before the auditing and broadcast deadline. Six days was not even sufficient for producing a piece of news, let alone a feature program which required innovations in content, shooting and production. What a challenge!

The first difficulty came from the shooting equipment. For shooting news, a Panasonic P2 video camera would be enough. But to shoot "Great National Craftsmen," we needed high-definition video cameras and corresponding equipment to produce the optimal effects in images, lighting, sound, etc. The equipment we adopted included: SONY F5, Canon 5 D3, Panasonic P2 high-definition camera, three tripods for the above, six camera lenses, long-line microphone, wireless chest microphone, sound console, three LED lamps, a head lamp, a monitor, photography orbit, two computers for editing (an Apple computer and an EDUS), hard disk, and so on. The back seats and trunk of our car were filled with so much equipment that the front of the car was raised. The cameraman and I were already exhausted by various chores such as borrowing and transportation of equipment, lighting, changing lenses, laying sliding rails, etc., not to mention the shooting.

There were three members in our team. One was the sound engineer. The cameraman Shao Chen and I were responsible for the rest of the work, including driving. When we arrived at Meng Jianfeng's company the first day, he was amazed as we took all the equipment out of the car.

As Meng Jianfeng's work did not involve much travel, it seemed that the shooting would be easy. However, it was not the case. Shao Chen and I live in Tongzhou District, and Meng Jianfeng's company is in Haidian District. The two places were far apart, and the traffic was very bad. Every day we spent more than four hours commuting. We left home at 6 a.m. and got back at 12 p.m. We had only two days for shooting, yet we had to copy the footage into the editing computers and send the same back to the CCTV station within the two days. When I got home at midnight, the first thing I did was turn on the computer and send the materials to the TV station. When I finished, it was already 2 a.m., yet I had to get up at 5 a.m. for the next day's shooting.

What tormented me was not only the intense work, but also the immense psychological pressure. Meng Jianfeng, with his passion for traditional Chinese craftsmanship, is constantly pursuing perfection. I have engaged in the profession of journalism for more than 20 years, and I too have been pursuing perfection and innovation. Therefore, the tight schedule and other disadvantageous factors could not serve as excuses. Both Shao Chen and I were determined to produce a high-quality program.

Although Meng Jianfeng works in a factory built in the 1980s, his work is quite demanding and does not admit any mistakes. Compared with other episodes of *Great National Craftsmen* in other fields, those in the aerospace field in particular, there were no imposing modern factory buildings or dazzling large mechanical equipment in our program. It could be said that the shooting environment was not ideal at all. With the dim light and monotonous and narrow space, the shooting of all scenes had to be well-conceived and sufficient light had to be arranged, or the scenes would be dark and dull. On the first day, Shao Chen worked under great pressure for nearly 20 hours. The next day, he felt dizzy, nauseous and even vomited under the physical and mental pressure. Yet he immediately resumed his work after a brief rest. He was quite strict about each shot, pursuing perfection. When the program was broadcast, the audience were deeply impressed with and amazed at the screen effects. The program well presents the exquisite craftsmanship of traditional Chinese chisel engraving as well as the figure of the great master Meng Jianfeng, who is dedicated to the craft.

Due to the tight schedule, I did the post-production work of scripting and editing, etc., two days after the shooting. Shao Chen continued with the interview and shooting all by himself in hot weather, and then sent the materials back to the TV station. With the help of my colleagues in the press release team, the post-production content was completed in two days. When we left the TV station around 4 a.m., it was still dark. I drove my colleague Kang Kang home. On the way, we saw how the city was being lit up by dawn bit by bit. When I got home, I even marveled at my own tenacity.

Xu Qiang, deputy director of the News Center, spoke highly of the program after reviewing it. I could have had a good rest, but I was still concerned about the program. As we had had a tight schedule, the editing of some shots was less than perfect. Although I was off-work the next day, I decided to make some improvement. I went to the new CCTV station to get the material, and sent it back to the old site. I worked on the modifications till 1 a.m. When I finished and got home, it was already 3 o'clock in the morning. But when the program was broadcast on the "Morning News" at eight o'clock, I felt that all my efforts had been worthwhile.

My colleagues and I spared no effort, partly driven by our professional ethics and partially because of the inspiration we had received from the hero of our program, the great craftsman Meng Jianfeng.

Before the shooting, I hadn't any idea about chisel engraving, nor did I know much about the arts and crafts trade in general. But in just two days, I learned from Meng Jianfeng not only about chisel carving, but also the essence of the "spirit of chisel engraving."

Meng Jianfeng is very stubborn in sticking to his own views. In the interview, I asked him several times if it was necessary to be concerned with a tiny, negligible flaw. Invariably, he replied with his favorite words – "perfection," "flawlessness" and "surpassing."

Meng Jianfeng's profession is different from those of others, who deal with such things as rockets, submersibles, large aircrafts and ships. For the latter, there are clear standards and objective measurements, for example, in the evaluation of precision level. As those fields concern life-and-death matters, and any failure means huge disasters and losses, absolute conformity to standards is required. Therefore, the sense of responsibility is an absolute requirement in those fields. In the arts and crafts trade, however, there are no objective standards, and variance in the artworks will generally not lead to serious results. Therefore, each artwork is a mirror of the skills as well as sense of responsibility of the craftsman. Meng Jianfeng is an excellent example in this regard. I had been searching for the essence of his spirit while we were making the program, and I got to know its value and greatness in the process. At the same time, Meng Jianfeng's story was a great driving force for me to come up with a high-quality program despite the tight schedule. I conquered and surpassed myself. Just as Meng Jianfeng is dedicated to creating national gifts and other exquisite artworks, we were also dedicated to perfecting the program about him with great responsibility and perseverance.

By LI XIN, reporter of China Media Group & CCTV News Center

GU QIULIANG

Fitter with a Precision of 0.02 mm

Profile

Gu Qiuliang used to be a senior technician of China Ship Scientific Research Center, China Shipbuilding Industry Corporation, and leader of the general assembly group of *Jiaolong*, China's first self-designed and self-developed manned deep-sea submersible vehicle (DSV). After his retirement in 2015, Gu Qiuliang was re-employed for the development of the latest "deep sea warrior" due to his rich experience and expertise, and was made responsible for the installation of the key parts. Because Gu Qiuliang can reach a precision of 0.02 mm by observing with the naked eye and touching without using any artificial instrument, he is called "Gu Liangsi" by his colleague ("Si" being a length unit that equals 0.01 mm, and "Liangsi" meaning 0.02 mm in Chinese).

Gu Qiuliang, senior technician of China Ship Scientific Research Center, China Shipbuilding Industry Corporation

Preview

A deep-sea manned submersible vehicle is composed of hundreds of thousands of parts, and the biggest challenge in its assembly comes from sealing, which requires a precision of 0.01 mm. In the assembly of DSVs in China, Gu Qiuliang is the only person who can reach this precision; therefore, he is known as "Gu Liangsi."

"Attention all posts, the submersible will now be deployed."

Liu Feng, in charge of the *Jiaolong* sea test site, gave the diving instruction. Gu Qiuliang was very familiar with the voice. Following the instruction, the *Jiaolong* was hoisted onto the No.09 Deck of the mother vessel *Xiangyanghong*, and deployed in deep water for yet another performance test.

In deep water, the pressure on an area the size of a fingernail is as much as 1 kg. To ensure the waterproofness of the submersible and resist huge water pressure, the assembly precision of all sealing surfaces of the manned submersible must reach the *si* level, or 1/10 of a human hair (1 *si* is 0.01 mm). Gu Qiuliang is the only person who can reach this precision in manned submersible assembly in China.

After the *Jiaolong* passed the sea test and was delivered for service, Gu Qiuliang was confronted with a new challenge – assembling China's first self-designed and manufactured 4,500-m-long manned submersible.

Gu Qiuliang was quite clear about the challenge: "The *Jiaolong*'s operations capsule was custom-made in Russia, but this one was made by ourselves. The difficulty of assembly was to keep the contact surface between the steel and glass parts below 0.2 *si*."

0.2 *si* is only 1/50 of a human hair. It may not be difficult to attain such precision with instruments, but the glass of the observation window of the manned capsule is too delicate for contact with any metal instrument. In the deep sea, where the pressure is several hundred atmospheres, any small scratch caused by friction with an instrument might make the glass leak or even break and endanger the lives of the crew. Therefore, the assembly of the capsule glass is the most demanding work in the assembly of a manned submersible.

Gu Qiuliang and his colleagues came up with a solution.

"The glass of the manned capsule is even softer than a camera lens. In fact, it is so soft that one can't even touch it with a fingernail. We must be extremely careful. When we assemble the glass, we lift it with rubber suckers. Then we hold it with our hands on the other side and wipe it clean."

For Gu Qiuliang, it is more important to rely on his own judgment, i.e., to observe with naked eyes and touch with his own hands, than to rely on instruments.

It is no exaggeration to say that Gu Qiuliang, with his naked eyes and hands, surpasses precision instruments. He can achieve a flatness within 2 *si* in the manual grinding of the seal surface of submersibles even in rough seas.

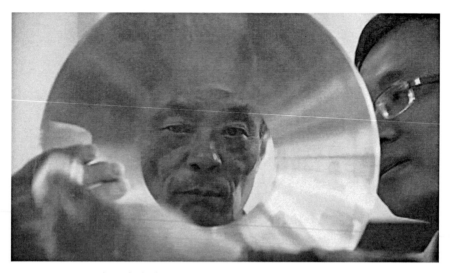

Gu Qiuliang checks the glass of the observation window of a manned capsule.

Gu Qiuliang, renowned and accomplished as he is today, had the experience of being much criticized by his masters when he was an apprentice.

Gu Qiuliang's second master, Zhang Guibao, criticized him severely: "Gu Qiuliang used to be rebellious and stubborn, so I often scolded him."

Today Gu Qiuliang has trained many apprentices himself, but he still appreciates the sternness of his masters. "Once I performed poorly. My master said that if I didn't take my work seriously, others wouldn't trust me. He said that I had performed so badly that he could not teach me anymore, and suggested that I learn from others. I was very sad."

Thanks to the strict training of his masters, Gu Qiuliang became more and more serious about his work. Finally, he mastered the basic skills with the utmost patience and efforts.

To practice basic skills, he was told to file a 10-cm steel cube into a 0.5-cm one. While he was filing, he kept summarizing the experience. After finishing about 15 steel cubes and breaking dozens of files, Gu Qiuliang became more and more proficient. The workpieces he made were so standard that no test was needed. Gradually, he was addressed as "Gu Liangsi" by his colleagues.

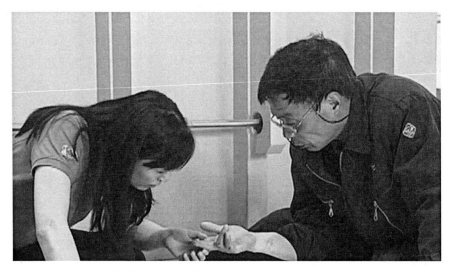

Gu Qiuliang shows how his fingerprints have worn away.

Gu Qiuliang sometimes jokes with his colleagues that his skills could be tested with a hair: "Just as a skilled runner can run with a bowl of water without spilling a single drop, a skilled fitter can file any surface flat. I can do this to a precision of 0.2 *si*. You can test it with hair if you don't believe it."

In 2004, Gu Qiuliang and some of his senior colleagues were selected to assemble the *Jiaolong*. As Gu Qiuliang was highly skilled, he was appointed the leader of the assembly team. Their biggest challenge was to ensure the water-tightness of the submersible.

Ye Cong, chief designer and first test leader of the *Jiaolong*, explained the importance of this operation: "Any defect would be disastrous. A pressure of dozens of atmospheres could tear apart a manned capsule."

As the *Jiaolong* was China's first manned deep-sea submersible vehicle, there was no precedent for Gu Qiuliang and his team to refer to. They had to experiment all by themselves: "It was like shadow boxing. I had to feel with my hands for long time to get the correct feelings. It was also like getting along with others, that is to say, affinity will grow after people spend enough time with each other. Look, some of my fingerprints are already worn away, and I have to use my ring finger to make a fingerprint."

When Gu Qiuliang first joined the *Jiaolong* project, the lab he had previously worked for wanted him to go back, offering him a higher salary. As Gu Qiuliang's wife didn't have a job and his daughter was attending school, the family was very much in need of money.

Gu Qiuliang's wife Wu Jingxia understood and supported him. She said, "We thought about the issue. My husband made his choice, for the project needed him more."

Apart from the offer of a higher salary, there was another thing that made Gu Qiuliang waver. He suffered from seasickness, and going to sea is a big physical challenge for him. "(In the first sea test of *Jiaolong*) I vomited whatever I ate. In a week I ate only a packet of instant noodles. I lost 13 kg in the three months at sea. I told my daughter jokingly that the best way to lose weight was to go to sea. If you suffer from seasickness, you will surely lose weight in one week."

However, Gu Qiuliang did not give up. "How could I leave when the underwater pilot trusted me with his life? The *Jiaolong* was like my child. I had to bring it up."

Gu Qiuliang stayed because of the great trust and responsibility. 3,000 m, 5,000 m, 7,000 m . . . As the *Jiaolong* went deeper and deeper, Gu Qiuliang had to make more and more technical breakthroughs and he had less time for his family. In the 7,000-m test, Gu Qiuliang received an urgent phone call from his wife, in whom a medical examination had suspected cancer.

When Wu Jingxia recalled those days, she couldn't help shedding tears. "It was terrible, but I had no one to turn to. What could I do? Then the labor union, the Party branch committee and some leaders of my husband's unit came to comfort and help me."

Nevertheless, after a hard mental struggle, Gu Qiuliang continued with his work on the sea.

Looking back on his 43 years of work, Gu Qiuliang feels he owes his family the most. He is 64 years old, having retired in 2015. Recently, he has been learning to drive. He plans to spend all his time with his family now.

"I want to go sightseeing with my family. Our country is so large and beautiful, and we want to see more of it," he said.

Currently, there are two manned deep-sea submersible vehicles in China, and Gu Qiuliang was in charge of the assembly work of both. The 4,500-m submersible might be the last submersible he has assembled. After the glass of the capsule was

The Jiaolong *descends.*

in place, Gu Qiuliang checked its safety repeatedly and carefully, as he has always done.

Gu Qiuliang defined his work as follows: "It is easy to be trusted once or twice, or for one or two years, but it is difficult to be trusted throughout life. Only when you are satisfied with yourself will others be satisfied with you."

The trial pilot Ye Cong is Gu Qiuliang's most intimate partner. In Ye Cong's eyes, Gu Qiuliang's standard operation is the guarantee of the safety of himself and other pilots. "Before we descend, Gu will pull out the safety pin and wave to us, which makes us greatly reassured, for we know that the submersible is safe."

In his 43-year career, Gu Qiuliang has been pursing lifetime trust with his faith, hard work, assiduous study and his efforts to challenge limits. Gu has earned the trust of the lives of the submersible pilots, and has witnessed China's progress from a country with only vast ocean resources to a maritime power.

The Jiaolong *returns to the sea surface.*

Great Craftsman Guarding the *Jiaolong*

Deep-sea scientific research capability is one of the important indices for measuring a country's comprehensive national strength. After continuous efforts over more than a decade, the *Jiaolong*, China's first self-developed manned submersible with a depth capacity of 7,062 m, has demonstrated China's deep-sea research capacity to the whole world. Throughout the process of its design to deep-sea operations, the patriotic and dedicated researchers and workers have been promoting the progress of the country with their expertise and hard work.

This year is the tenth since I got to know Gu Qiuliang.

In April 2010, the *Great National Craftsmen* series came into being with the implementation of the manufacturing power strategy. Perhaps the initiator of the program did not expect that CCTV's own brand "Craftsmen" would last till today and will continue in the future. Our era calls for craftsmen! Until May Day this year, the "Craftsmen" series has been broadcast for six seasons. I am so lucky to have been able to participate in the previous five seasons and documentaries. Gu Qiuliang was the craftsman I chose to report on in the first season.

In April 2015 the team for "*Great National Craftsmen*" was established. Jiang Qiudi, then deputy director of the Economic News Department of CCTV's News Center brought us together. Mr. Yue Qun was the leading producer of the program. Our task was to select a person who deserved the title of great craftsman, and make the first episode. At that time, the *Jiaolong* manned submersible had completed sea tests and been delivered to the national deep-sea base. As a reporter who had followed the whole process of the *Jiaolong* from project approval to research and development and then to sea experiment and subsequent experimental application, I had formed a close relationship with the *Jiaolong* team. Over more than 10 years, I grew familiar with every post and job. China's manned deep-sea diving project is a key national 863 project. There must be more than one who deserved the title of Great National Craftsmen in the *Jiaolong* team.

I visited the China Ship Scientific Research Center of the China Shipbuilding Industry Corporation, the designer and manufacturer of the *Jiaolong*. The Center recommended several candidates to me, and Gu Qiuliang was one of them.

When I heard Gu Qiuliang's name, the following scene immediately appeared in my mind: A man was chewing his food slowly in a canteen. His eyes squinted, and he seemed somehow dizzy, apparently from seasickness. I was most impressed by the scene, because both he and I suffered from acute seasickness. According to seamen on board, one would no longer suffer from seasickness after one endured it for a week. However, we were not so fortunate; seasickness accompanied us throughout the entire voyage. Because of the shared experience, I got familiar with Gu Qiuliang, and we soon became friends.

Even so, the interview didn't go well. Sitting in the conference room of the China Ship Scientific Research Center, Gu was obviously not as at ease as he was when he was at sea. What did he usually do when he returned from the sea? Obviously a face-to-face conversation wasn't enough to get a full glimpse of his life. So we changed our strategy. For the next five or six days, we followed Gu Qiuliang from Wuxi to Luoyang, spending as much time as possible with him every day.

I also talked a lot with his colleagues. His colleague Ye Cong described his superb skills to us. "He can basically judge the flatness of a polished surface without instruments. His polishing accuracy can reach 2 *si* (1 *si* is 0.01 mm). Therefore, we call him 'Gu Liangsi.'"

Si is a measurement unit of length equal to one tenth of a normal human hair. Gu Qiuliang's top skill is to control the sealing degree, i.e., the gap between the metal chassis and the glass of the manned capsule of a submersible within 0.2 *si*, so as to ensure zero leakage of the submersible at depth as well as to avoid the glass breaking due to different water pressure resistance capacities of glass and metal.

Si would be a highlight of our program. As the thought flashed through my mind, I began to observe carefully every movement of Gu Qiuliang, i.e., how he observed workpieces and the movement of his fingers in working, how he took a rest, and so on. While Gu Qiuliang was working, he had two habitual motions: touching and observing. He would touch the field steel plate, the metal surface of the cabin window, the glass mounting surface, and so on. Before touching, he would invariably observe. Gu Qiuliang was really good at touching and observation. In order to test his expertise, we proposed jokingly to test it with a hair. I pulled out a fine hair, and gave it to Gu Qiuliang. Gu fixed his eyes and fingers on it and said that the hair was no more than 2 *si*. Then he measured it with a vernier caliper, which proved him correct.

This detail became a highlight of the episode. During the review of the program, Xu Qiang, duty director of the CCTV News Center, specially asked us to zoom in on this detail. In retrospect, I am still amazed at Gu Qiuliang's exceptional skill. I am also deeply touched by his perseverance in practicing his skills. In the past 40 years, he has felt so many workpieces that his fingerprints are practically worn off. In the days of the interviews, Gu Qiuliang was learning to drive when he had time. However, he was unable to submit his fingerprints for the license. Repeating a seemingly common operation for more than 40 years, this is the spirit of a great craftsman.

Gu Qiuliang likes to smile. When I interviewed him in 2015, he told me with a smile that he would retire in the same year. He would learn to drive, and take his wife to travel around the country and the world after retirement. Because he had been preoccupied with the development and sea experiments of the *Jiaolong* for years, he hadn't had much time for his family. He likes to smile, which doesn't mean there aren't bitter moments in his life. Years ago, he participated in the 7,000-m sea test of the *Jiaolong*. As soon as he set off, his wife called him, saying that she might have cancer. He was the backbone of the family, but the sea test was also essential to the success of the *Jiaolong*. Everyone on board was indispensable. He had the chance to return, but he decided to stay with his team. It was a tough decision. The *Jiaolong* was like his child. In order to guard it well, he would rather bear the misunderstanding and even complaint of his family. What was more, inside the manned capsule were his team brothers, with whom he had shared weal and woe for so many years. It was his duty to guard their lives. Fortunately, his wife's call was a false alarm, i.e., her tumor was benign. After Gu Qiuliang returned, he and his wife hugged each other, and cried heartily. He was both ashamed of his absence and rejoiced that his wife had escaped cancer.

Over the years, all the members of the *Jiaolong* team went through thick and thin together. Looking back on those days, everyone is still deeply touched. How many times have they observed the data transmitted from the *Jiaolong* deep under the sea to the on-site headquarters of *Xiangyanghong 09*? How many times have they waited for the submersible's return? How many times have they worked overnight troubleshooting when the test data turned out unsatisfactory? How many times has Gu Qiuliang pulled out the safety pin of the *Jiaolong*, signaling that the cabin was safe? . . .

50 m, 1,000 m, 3,000 m, 5,000 m, 7,000 m . . . As the *Jiaolong* descends deeper and deeper, its ability of deep-sea exploration is gradually enhanced. Gu Qiuliang, a member of the *Jiaolong* team, is a great craftsman. Thanks to these responsible and dedicated front-line workers, China is becoming a manufacturing power.

It is easy to win others' trust once or twice, or for one or two years, but it is difficult to earn lifetime trust. Over the 43 years of his career, driven by his firm conviction, Gu Qiuliang has been dedicated to hard work and assiduous study. He has been challenging the limits and pursuing trust throughout his life. With a firm conviction, he has won the utmost trust of the submersible's crews, and also witnessed China's progress from a country with only vast ocean resources to a maritime power.

By Wang Kaibo, reporter of
China Media Group & CCTV News Center

ZHOU DONGHONG

Xuan Paper Master

Profile

Zhou Donghong, born in 1967, is a senior technician of China Xuan Paper Co., Ltd.

Zhou joined Jingxian Xuan Paper Mill, Anhui Province, in 1985. Ever since then, he has been engaged in Xuan paper production. He holds the zero-defect record for finished products. In the past 30 years, he has inherited and developed traditional skills, improved the paper-making process and trained more than 20 leading technicians. He has contributed much to the inheritance of Xuan paper-making craftsmanship. He has made commemorative Xuan paper for the 60th anniversary of the founding of the People's Republic of China, for the craft's nomination as an intangible cultural heritage, and for the return of Hong Kong.

Zhou Donghong also takes an active part in social benefit activities. He has contributed money 40 times and supplies 30 times, the total value of which has exceeded 50,000 yuan. He has been awarded the honorary titles of "Exemplary Production Worker" and "Excellent Employee" more than 20 times. In 2015, he was selected as a "National Model Worker" as well as one of the top-eight "National Craftsmen."

Zhou Donghong, senior technician of China Xuan Paper Co., Ltd.

Preview

Zhou Donghong has been making Xuan paper for 30 years. He is totally devoted to this career, which is his lifelong pursuit. He studies the relevant skills and pursues perfection with unremitting perseverance. Every piece of his Xuan paper is a legend, and so is he.

In ancient times, the natural beauty of southern Anhui Province attracted many Chinese literati, such as Xie Tiao, Li Bai, Bai Juyi, Han Yu, Wang Wei, Meng Haoran, Li Shangyin, Su Dongpo, and so on. Some of them, such as Mei Yaochen and Hu Shi were native Xuancheng residents. Xuan paper ranks the first of the "four treasures of the study." As Xuan paper has been named a world intangible cultural heritage, the craftsmanship of making Xuan paper has attracted increasing attention worldwide. Li Keran, a famous Chinese painter, once said, "Without good Xuan paper, there will not be good Chinese paintings that can be passed down through the ages." The production of a piece of Xuan paper consists of more than 100 procedures. In these procedures, paper leaching determines whether the process will succeed or fail.

Xuancheng is the birthplace of Xuan paper. Thanks to today's high-speed rail, it takes only a little more than five hours from Beijing to Xuancheng. Jingxian Xuan Paper Mill is located by Taohuatan ("Peach Blossom Pool"), where the great poet Li Bai of the Tang Dynasty once bade farewell to his bosom friend Wang Lun.

It is 2 o'clock in the morning. In the workshop of Jingxian Xuan Paper Mill, Zhou Donghong and his partner are busy with the "paper leaching" process. Standing face to face, they hold a sheet of paper, and sway it from side to side in a sink, and a piece of wet Xuan paper takes on its preliminary look. Although the procedure takes only a dozen seconds, it determines the thickness and texture – and therefore the quality – of the Xuan paper.

In the leaching process, there is a ten-character formula, the main idea being deep leaching first and shallow leaching to follow. In the deep leaching, the operators hold one side of the leaching screen with their hands above the paper pulp sink and the other side at shoulder height by lifting their other hands to about 45 degrees. Then they lower the screen into the sink to have the pulp cover the screen completely and evenly. In shallow leaching, both operators put the screen into the sink about 15 cm deep with their hands at an even level, when the pulp on the screen becomes more even and smooth under the force of the water in the sink.

Only experienced workers like Zhou Donghong know how hard the job is. The ink permeability feature of Xuan paper poses strict demands on the thickness of the paper, which requires the operator to be steady and exact with each movement. Despite the pulp concentration reduction in the paper-making process, the

Zhou Donghong is very serious and meticulous in his work.

Zhou Donghong repeats the paper-leaching operation over 1,000 times a day.

thickness of all paper made should be almost the same. Moreover, as the mill produces more than 100 types of paper, the varied thickness of the products has to be taken into consideration.

The leaching work often starts at 1 a.m. and continues till the evening. Every day, Zhou Donghong and his partner repeat the operation more than 1,000 times. His requirement for himself is that the weight error range of one *dao* (*dao* being the quantity unit for Xuan paper, with 1 *dao* being 100 sheets) of Xuan paper should be within 50 g. This is to say, the weight error of each sheet should not exceed 1 g.

Zhou Donghong is quite confident about his skills. "The 50 g criterion is essential in determining the quality of Xuan paper. If it is thinner or thicker, it will fail to reach the best ink permeability effect when a calligrapher writes on it. My hands are like scales. Over the past 30 years, the weight error range of each *dao* of Xuan paper I produced never exceeded 50 g."

Zhou Donghong is highly proficient at the leaching process, and each of his movements is smooth and natural. However, he almost gave up when he began to learn the trade years ago. Paper leaching is conducted through the cooperation of two operators, one performing the main control role and other the auxiliary role. The main operator earns 20% more than the auxiliary operator. When Zhou was young, he was self-confident, and wanted to be the main operator. However, it was quite difficult to master the gist of the seemingly simple jargon of deep leaching first and shallow leaching to follow. Zhong worked hard for long hours, but he failed to complete his workload in the first month. As some of his colleagues had been doing the auxiliary job for a dozen or even dozens of years, Zhou Donghong was painfully frustrated.

However, Zhou Donghong cared greatly about face-saving. He used to be a farmer, and it took him much effort to become a worker in a state-owned enterprise. In the eyes of his relatives and friends, he was a promising man. What was more, his mother, an otherwise gentle and loving woman, firmly opposed his idea of quitting his new job, for it would be a huge waste and humiliation to give up. She even claimed that she would stop feeding him. Zhou Donghong went back to the paper mill because of his mother's tough attitude. "Since I have to do it, I will do it well!" he promised himself from then on. He settled down to learn from his experienced colleagues, and practiced diligently.

In Jingxian Xuan Paper Mill, systematic training was held for new apprentices, which gave Zhou Donghong a second chance to learn paper leaching. Every day, Zhou came to the factory early in the morning and worked hard all day. He even kept thinking about paper leaching when he was off work. While his colleagues were working, he would observe closely every detail of their operation, and came to understand the subtle changes. He would ask when he had questions, but more often he would solve problems by himself.

Early one morning, Zhou Donghong had worked for several hours, when he suddenly felt that the weight of the piece of paper he had just leached was different from those he made when it was dark. He went to ask his supervisor, the best paper-leaching worker in the factory. His supervisor was very glad at his discovery. He explained that the reflection on the paper-leaching screen was different when it was dark and when it was dawn, which would affect the judgement of the operator. From then on, the supervisor knew that Zhou Donghong, who used to be paid less than 20 yuan a month, was finally approaching the correct way with his own efforts.

Diligent study and strenuous practice meant that the monotonous paper-leaching operation has become the most important part of Zhou Donghong's life. It was all right in summer, but in the winter Zhou Donghong developed painful chillblains from dipping his hands into the freezing cold water in the sink. Even so, Zhou Donghong did not stop working, because as long as he stopped for a few days, he would lose the feeling in his hands that he had accumulated after long-time practice. Zhou Donghong said that doing a good job of paper leaching required both keen understanding and hard practice. Finally, his efforts paid off, and he became a qualified worker.

There are more than 1,000 employees at China Xuan Paper Co., Ltd., but Zhou Donghong is the only one who can make the thinnest Xuan paper (138 cm × 70 cm), which weighs 1.4 kg per *dao* (100 pieces) and the thickest paper (138 cm × 70 cm) which weighs 5.7 kg per *dao*.

"Our company produces more than 100 kinds of Xuan paper, which requires workers to master the corresponding skills, get familiar with the composition of more than 100 kinds of water pulp slurry, acquire the corresponding hand feelings and appreciate the subtle differences..." Zhou Donghong, an otherwise taciturn man, will utter an unceasing flow of words when it comes to his paper-leaching

Zhou Donghong and his wife

job. His words are full of professional pride and confidence. "Many people ask me to make Xuan paper specially for them. Isn't that an honor?"

Year after year, Zhou Donghong often gets up in the middle of the night to do his job. At work, he often stands by the sink for more than ten hours. His wife has some complaints: "We are a midnight couple. He often leaves home at 1 or 2 in the morning, and works until 5 or 6 in the afternoon. Sometimes I wake up at midnight and find he has already left. He is concerned only with his job. Sometimes I ask him why he never says 'I love you' to me as couples on TV do. He replies that TV series are not real life, that he does love me but he won't speak out."

Although Zhou Donghong is preoccupied with his work and does not speak sugared words, his wife, Zhou's childhood sweetheart who works in the same paper mill, understands and supports him. She knows that after so many years, paper pulp leaching is not merely a means to make a living for Zhou Donghong; the

Xuan paper has long been a part of his life which he cherishes so much. As a sharp-eyed quality inspector, she can tell the mood of the leaching operators from every tiny flaw in the paper.

Zhou Donghong is 50 years old. Sometimes his wife asks him jokingly how long he intends to continue with his job. "Will you be leaching paper pulp until you retire at 60?" For Zhou Donghong, this is no problem at all. He replies that he likes paper pulp leaching, and will not cease as long as his health permits it.

Zhou Donghong is quite satisfied with his current situation. "If I had changed my job, I wouldn't necessarily end up with the achievements I have made today. There is an old saying in China that there are outstanding persons in every trade. I have hung on till today, and I'm very proud of myself." When Zhou said these words, his face glowed with pride.

Zhou Donghong said that Xuan paper was a legacy from our ancestors with a history of more than 1,500 years. The production of Xuan paper takes more than 300 days, 18 links and more than 100 procedures. Now he is beginning to think more about how to pass on his craftsmanship to future generations.

Zhou Donghong mixes paper pulp.

In recent years, Xuan paper production has been encountering a strong impact on the industry and severe market shrinkage. Because the production cycle of hand-made traditional Xuan paper is as long as three years, the production cost always remains high. Besides, a lot of other calligraphy and painting paper has gradually occupied the medium and low-end markets with ever-reducing cost, some of which even go under the name of "Xuan paper." Moreover, qualified Xuan paper workers like Zhou Donghong are getting older and older, and fewer and fewer young people are willing to learn this trade.

Zhou Donghong used to have a dozen apprentices, but only half of them chose to stay. Many young people leave because the seemingly simple operations require practice over a long time, the work is strenuous and boring, and the humid working environment causes occupational diseases. Zhou's first apprentice, Zhao Zhigang, who is also his favorite one, is 47 years old. In 2012, a college graduate applied to learn the trade. The factory thought highly of him, and offered him a starting salary 1,000 yuan higher than normal after he became a regular employee. Nevertheless, the college graduate left after two years. Zhou was disappointed but could do nothing about it.

The shrinking of the market is leading to a loss of talented people for the industry. The intangible cultural heritage of Xuan paper making is facing the risk of extinction, too. Currently, the average age of workers in the factory is over 40, and the mobility rate is higher than before. Zhou Donghong is worried that there will not be enough young people to inherit the craft. Fortunately, more and more people are supporting him now.

In 2015, Zhou Donghong received a National Labor Medal for the first time. In the same year, the production process of the paper mill was also improved. Zhong and his colleagues no longer need to start working early in the morning. Now they work eight hours a day and five days a week. At the end of that year, the China Xuan Paper Museum was established near the Xuan Paper Cultural Park in Jingxian County, and Xuan paper culture was included in the local tourism projects.

Today, Zhou Donghong engages in not only Xuan paper making, but also the promotion of Xuan paper culture. He is an ideal person for the job. In the eyes of his colleagues, after receiving the National Labor Medal, he no longer wears as casual as before. Since he is now a spokesperson for Xuan paper culture, increasing tourists want to take photos with him.

In the summer of 2016, Zhou Donghong participated in the "Knowledge Training on Folk Culture & Intangible Cultural Heritage Projects" at Tsinghua University in Beijing at the invitation of the Ministry of Culture. The training course lasted over a month. He communicated with intangible culture masters from all over the country, and promoted Xuan paper culture.

Although he is now a renowned figure in the Xuan paper-making business, he is as hard-working as ever. He still does his paper pulp leaching job for long hours every day. He said only in this way could he keep the feeling in his hands.

Zhou Donghong said that he did not know what the spirit of a craftsman was, but he knew that one must study and work hard to do one's job well. This is his life-long conviction. In his more than 30 years in the industry he has made nearly ten million sheets of Xuan paper, all of them first-class.

Remain true to one's original aspiration, and one will accomplish one's mission, Zhou Donghong stresses. He always strives for perfection in the traditional craftsmanship of Xuan paper making. In the process, he experiences the joy of working as well as the pride of inheriting one of mankind's intangible cultural heritages.

Zhou Donghong in Real Life

Interviewing the "Xuan paper making master" Zhou Donghong was not our original plan. However, it happened and turned out quite successful. We had planned to interview two figures in the construction industry, but we found that it was difficult to show their work in TV programs. After that, I contacted masters in traditional and modern manufacturing industries, but I failed to spot the appropriate filming subject. After all, there are not many who deserve the title of "Great National Craftsmen."

I live near Liulichang, a collection and distribution center of traditional handicrafts, antique calligraphy and paintings, etc., in Beijing. Goods sold there also include the "four treasures of the study" (namely writing brush, ink stick, ink slab and paper). Along both sides of this street of a few hundred meters, there are many shops selling Xuan paper. Xuan paper is generally expensive. According to shop owners, Xuan paper is hand-made through very complex processes, and the most difficult part is to make it even and of uniform thickness. I immediately got an idea: There must be great craftsman in this trade! Moreover, the elements of Xuan paper, the "four treasures of the study" and traditional crafts best represent Chinese culture. So I called a friend at Anhui TV Station, and asked him if there was such a person who could be interviewed and filmed.

We started without delay. On the 19th, we decided upon Xuan paper as the theme of our program. On the 22nd, we arrived at Jingxian County, Anhui Province, the most authentic place of origin of Xuan paper. With the coordination of my peers at Anhui TV Station, the local government and the Xuan paper factory attached great importance to our program. They recommended Luo Ming, a provincial intangible cultural heritage inheritor to us. Luo Ming is the deputy general manager of China Xuan Paper Co., Ltd. He introduced the history, process and quality standards of Xuan paper to us in detail, but he also admitted that although he was the inheritor of Xuan paper making, he had long quit production to engage in management. He recommended another person, i.e., Zhou Donghong, to us. Zhou had engaged in paper pulp leaching work for 30 years, and he is a national model worker.

Interviewing Zhou Donghong was not an easy job. The first interview was held in the paper pulp leaching workshop. Apart from introducing the paper pulp leaching process, he would only say one thing: "I was born for the job." We filmed him while he was working, and we soon found that the process of paper pulp leaching seemed quite simple. In our eyes, the secret of "deep leaching first and shallow leaching to follow" was no more than putting the screen into the sink and shaking it once. Two workers are required for the operation. Why could some workers only assume the auxiliary role after a dozen years, but Zhou Donghong was able to assume the main role in two years? There must be some secret. But Zhou insisted that he was born for the job, and mastered the skill quite naturally. When asked repeatedly, he replied with "Diligent study and hard practice." When asked for more details, he said nothing but "Just practice like this."

Zhou Donghong was busy during the days of our interview, because he had been selected as national model worker of the year, and had to attend various report and commendation meetings. The first interview ended hurriedly, as he had to attend a celebration dinner with local leaders that evening. The next day we went to his home again in the hope of getting to know more about him.

We talked about his life, his family and his daughter. He said he was satisfied with his life. As a return from years of work, he has received various honors and a considerable income. Our conversation lasted more than an hour. When I asked him how he had mastered his skills, I hoped very much that he could utter some high-profile and inspirational words, but he just said proudly that he was born for it. In my opinion, Zhou Donghong is a man of great determination who is proud of his profession. This is the temperament that a top craftsman must have. That evening, I also interviewed his wife. It was more like a casual chat. She told me that they were "a midnight couple." She complained jokingly that her husband never said "I love you" to her, mentioned his lifelong enthusiasm for his job, and so on. I think his wife's words lent more reality to Zhou Donghong's image in our program. He was like an ordinary person in our neighborhood, so real and approachable. Actually, the interview with Zhou Donghong's wife was my favorite part of the program. After the interview, we worked until half past one in the morning of the following day. The day after the interview, Zhou Donghong was to attend a national model worker commendation conference in Beijing.

The third interview took place after Zhou Donghong returned from the commendation conference. I called and asked him about his feelings on his Beijing trip. This time, we talked for about an hour again. During the conversation, I was happy that he recalled two details about when he was learning the trade years ago. At the beginning, he had no idea about the weight of paper pulp on the screen. Once, he dropped a bit of the pulp when lifting the screen. The factory director saw it, and said to him: "Zhou, this bit of paper pulp was more expensive than lard." This was a serious and effective reminder, for life in those days was so hard that lard was a luxury. Another time, Zhou Donghong felt that the weight of the paper he had leached during the day was different from that at night. He consulted his supervisor, who told him that the reflection on the paper leaching screen in the day was different from the same thing at night, which would affect the judgement of the operator. Zhou Donghong said: "In order to develop the correct feeling, I must leach the paper pulp every day, or I will lose the feeling. This is what diligence means. I keep doing it every day, and it is hard work. But only in this way can I maintain the correct feeling." On hearing these words, I was very excited. Being dedicated to one thing for 30 years with a zero-defect rate of finished products exactly revealed the spirit of the craftsman!

If the hustle and bustle of modern society is like a big river running forward in roaring, then the inheritance of Xuan paper craft is like a small stream that always flows slowly and steadily. The spirit of the Xuan paper craftsmen is revealed in this contrast. Zhou Donghong said that in order to maintain the hand feeling, he must leach the paper pulp every day. He likes this seemingly monotonous work. He doesn't know how to define the spirit of a craftsman, but he knows that he must endure loneliness, experience carefully the subtle difference of the repeated operations and achieve a zero-defect rate with the utmost efforts. This is Zhou Donghong's original aspiration. The essence of the spirit of a craftsman lies in diligent study and practice, and unremitting exploration, as well as the reverence and sense of honor one holds towards one's job. With this spirit, each of us can become a great craftsman.

By Lu Wu, former reporter of
CCTV News Center

HU SHUANGQIAN

Craftsman in the Aviation Field

Profile

Hu Shuangqian was born in Shanghai on July 16, 1960. He entered the Shanghai Aircraft Manufacturing Technical School in October 1978. Two years later, he graduated and entered the CNC machining workshop of Shanghai Aircraft Manufacturing Factory (today's Shanghai Aircraft Manufacturing Co., Ltd.). He has worked as a fitter there for 38 years.

At Shanghai Aircraft Manufacturing Co., Ltd., Hu Shuangqian has participated in the manufacturing of the Yun-10, MD-82, Boeing B-737, Boeing B-787, ARJ-21 and C919. He has made outstanding achievements in technological innovation and technological transformation in titanium alloy processing and other procedures. He is the author of *Eight Major Methods for Standard Bench Operations,* which is known as "The Eight Major Methods" and "Hu Shuangqian standard working method" in the company. The Eight Major Methods are promoted in all the workshops of Commercial Aircraft Corporation of China Ltd.

Hu Shuangqian has won many awards and honors, such as Shanghai Quality Gold Award in 2002, National Model Worker in 2015, National Moral Model in 2015 and nomination for the second China Quality Award in 2016.

Hu Shuangqian, fitter of Shanghai Aircraft Manufacturing Co., Ltd.

Preview

The independent research and development of large passenger aircraft, which is at the top of the modern industrial system, fully reflect the comprehensive strength of a country. In this industry, hand workers are becoming fewer, but they are increasingly irreplaceable. For example, Boeing and Airbus, despite their high degree of automation, still keep hand workers in their global factories for the most demanding manual calibration and revision in their aviation production process. These are very precious talents.

In China's large aircraft industry there is such a hand worker – senior fitter Hu Shuangqian. He is an amiable, easygoing and neat man. He is of medium height, with short gray hair and a light complexion. For 40 years, he has processed millions of aircraft parts, with not a single defect.

In the Pudong area of Shanghai there are many modern factory buildings busy day and night.

This is the Pudong base of Commercial Aircraft Corporation of China Ltd. It is in the forefront of China's civil aviation industry and the cradle of China's civil aircraft.

The C919, China's first large jet trunkliner with complete intellectual property rights, is assembled in the component assembly workshop at the Pudong base, where the wings, cabin and other key parts of the aircraft are assembled. Then the assembled parts are delivered to the general assembly shop to become a complete aircraft.

The component assembly process involves more than 1 million nonstandard parts, 80% of which are designed and produced in China for the first time. In this process, any deviation from the standard of any part will pose a potential risk to the C919.

Most of these nonstandard parts are produced by Shanghai Aircraft Manufacturing Co., Ltd., a company with a history of more than 50 years, and COMAC's production base in Shanghai's Dachang District.

Hu Shuangqian has worked for many years for Shanghai Aircraft Manufacturing Co., Ltd. The previous company, Shanghai Aircraft Manufacturing Factory ("SAMF"), was established in 1950, and mainly engaged in the repair of domestic aircraft. China's first large aircraft was produced by Shanghai Aircraft Manufacturing Co., Ltd. The company is the assembly site of China's first international aviation manufacturing cooperation project as well as its international joint production workshop. It is the veritable frontline of China's civil aviation industry.

Hu Shuangqian has spent almost his entire life processing aircraft parts. At this moment, he is carefully polishing a precision part. He is wearing safety goggles and a dust mask. Surrounding him is the low humming of drilling. After the polishing, the somewhat rough surface of the aluminum alloy takes on a new shiny metallic color.

In the workshop where Hu Shuangqian works, more than 90% of the space is occupied by modern precision CNC lathes. Hu Shuangqian and his team work on several workbenches in a corner of the workshop. As Hu Shuangqian is a hand worker, he seems a bit "old-fashioned" here, just like the old-fashioned tools he uses.

Hu Shuangqian works on an aircraft part.

"This file is specially designed for processing titanium alloy. It's relatively expensive. These files over here are more common," Hu Shuangqian says as he introduces the various files in a drawer to us.

Hu Shuangqian, born in 1960, is approaching retirement. He is now the oldest bench worker in his company. As the leader of the bench worker team, he has trained almost all the other members of the team.

Most people think that aircraft are completely produced on assembly lines, but it is no so in fact. In this factory, with an area of 3,000 m², there is the most advanced production equipment as well as bench workers who work manually. It is a sharp contrast, but there is smooth cooperation between them.

"For example, it is more efficient to fix a small right angle on a part manually than by using a machine. Drilling urgently needed parts is the same case, because programing is required in drilling with a CNC lathe, while the same might be done manually in a shorter time." For Hu Shuangqian and his colleagues, it is quite common to grind and fix the edges and corners of parts of complex structures until they are flawless. Although Hu's role is irreplaceable, he adopts very low profile.

Of all procedures in aerocraft manufacturing, Hu Shuangqian's work is the most challenging.

In aerocraft manufacturing, tolerance is allowed, but the standards are quite strict and even harsh. For example, the nonstandard parts for C919 Hu processes must reach a precision level of *si*.

Si is a unit of length equal to 0.01 mm, or half of the diameter of a human hair.

"In terms of my hand grip, it is a little tight, but not too tight when a bolt is placed into a nut." Hu Shuangqian tries to describe the "hand feeling" he has developed after years of experience. To measure this tolerance, repeated measuring with precision instruments is required.

To better illustrate his point, he made demonstrations with his hands, which already had an indelible metallic color from long exposure to paint and aluminum.

In the past 40 years, the owner of the hands has processed millions of flawless parts, all according to strict criteria.

"Our work is a life-and-death matter, so it is different from most other jobs." Hu Shuangqian paused after he said this, his expression being quite serious.

There have been heated discussions on the quality and safety level of large domestic aircraft, but only the insiders know the real situation.

In China, in order to put an airplane into use, permission from the Civil Aviation Administration of China is required. To get this permission, the airplane must pass a review by the Federal Aviation Administration (FAA). The FAA is the most experienced and demanding authority in the industry. The review involves every step in the full life cycle of the aircraft, including design, manufacturing, assembly, test flight, mass production and flight. In order to ensure the safety of the aircraft throughout out its life, there are corresponding requirements on the fatigue strength and life span of even the smallest screw. There is a lifetime responsibility system for examiners who sign the final review report.

This "ultimate test" in the civil aviation field is called airworthiness certification. The large aircraft C919 is undergoing such a test. Only by scoring full marks in this test will the C919 be qualified to enter commercial operation.

If the parts Hu Shuangqian makes can meet the requirements without any reworking, the whole plane might come into operation earlier. This is all that he can do, and he must do it well.

As a professional worker in the industry, Hu Shuangqian has a strong sense of mission. To this day, few people in the forefront of China's civil aviation industry are more authoritative than he is.

Hu Shuangqian became interested in airplanes when he was a child. In 1980 he graduated from a technical school and entered Shanghai Aircraft Manufacturing Factory, as he wished, to be in time to witness an important milestone in China's civil aviation, i.e., the maiden flight of the Yun-10. When he was an apprentice in the technical school, he accompanied his supervisor in processing a few small parts for the Yun-10. Hu said that when he was new at SMAF he was quite thrilled to see the Yun-10 take off on its maiden flight. He said he would always remember the scene.

However, no one knew that something unexpected would happen.

Two years after the Yun-10's first flight, the development of the large aircraft was suspended. In 1986, due to various reasons, including its failure to meet the requirements of airworthiness certification, the Yun-10 project failed to enter the civil aviation market. The development of the Yun-10 was declared completely terminated. At that time, the large number of top technicians assembled for developing the Yun-10 suddenly faced the embarrassing situation of having no work to do. Although the SAMF was constantly striving for other orders, it had a poor performance and could offer workers low wages while other enterprises in Shanghai enjoyed thriving business, and commerce was growing rapidly, too. Many senior workers left the factory for other jobs.

"Many of my colleagues left. At the most difficult time, our salary was less than 100 yuan per month, whereas that of other factories was 300 or 400 yuan. One man in our workshop specialized in numerical-control programming. Someone offered him a salary of 3,000 yuan, and so he left." Talking about this, Hu Shuangqian had complicated feelings. He said that at that time, many local enterprises of major industries looked for technicians they needed at SAMF. Some even turned up in cars, so that they could take those who were willing to leave immediately to their factories. People might leave any day. Hu received an offer from a private company which would have tripled his salary.

He declined the offer, not because he wasn't short of money, but because he believed China would build its own aircraft anyway. This was important, and the

A statue in front of the Yun-10 on the lawn of the SAMF with the inscription "Never give up"

country would need his factory and himself. "I believed that my factory might still manufacture planes in the future. I didn't think it would go out of business."

"Our country would definitely support the aircraft project and our factory would not be closed down for sure." Hu Shuangqian said. Although he was optimistic about the future, he didn't know how long he could wait.

After the Yun-10 project was terminated, China's last complete Yun-10 aircraft was delivered back to Shanghai and placed on a lawn at the SAMF, its birthplace. The aircraft is open to the public. It is a 124-seat airliner with six seats in each row. It is spacious inside, with all the instrumentation, seats and interior decorations intact. When Hu has time, he often gets into the plane, looks around and stays inside it for a while.

In the following 20 years, Hu Shuangqian worked on OEM projects, manufactured aircraft parts for Boeing and McDonnell Douglas, and worked to produce China's first regional jet aircraft – the ARJ21. He worked on manufacturing airplanes, but they were not the big aircraft that he dreamed of.

In 2006, the C919 project, China's new-generation jumbo jet, was officially approved. It was China's first civil airliner with independent intellectual property rights and conforming to the latest international airworthiness standards. It is one of the 16 major projects identified in *The Outline of the National Program for Medium and Long-Term Scientific and Technological Development* (2006–2020). The Chinese dream of big aircraft was launched again.

Naturally, the SAMF became the front line of domestic civil aircraft manufacturing again. The new-generation C919 adopted a lot of brand-new designs, which meant higher requirements for the SAMF. Hu had to process spare parts manually and with maximum accuracy. The parts were of various sizes, some being nearly 5 meters and others smaller than a paper clip.

"The smallest part I made was about half the size of my little finger nail. It was like a hook, very small, and had holes in it." Hu Shuangqian said. He showed us a sample of the tiny hook. It looked a bit like the numeral 6. When we finally got our camera focused on the small metal hook, Hu Shuangqian's fingerprints were clearly seen on it, too.

During the interview, Hu Shuangqian was given another task: to process a piece of alloy according to drawings. He started working immediately. He turned on the computer and compared the piece of alloy to be processed with the electronic drawings repeatedly. He painted it, and marked it with a special knife for drilling and polishing. Then he drilled the alloy. Some small silvery metal pieces flew out, making the piece look like a blooming metal flower.

This is the routine work of Hu Shuangqian. He also deals with emergencies in the production process, which is far more challenging.

Once, the factory needed a special metal part urgently to assemble in a plane, but it would take several days to have it delivered from the manufacturing plant. In order to meet the schedule, the factory decided to process the special part with titanium alloy on site. This task was given to Hu Shuangqian.

"This titanium alloy blank was only the size of the palm of a hand, yet it was worth more than 1 million yuan. It was very expensive because it needed precision forging – 36 holes of different sizes had to be drilled in it, with a precision of 0.024 mm." Hu Shuangqian remembered the job clearly.

Precision of 0.024 mm is beyond the capability of the naked eye. As no drawing was available, it was impossible to process the part on the CNC lathe by fine programming. Hu Shuangqian had to use his hands and a traditional milling and drilling machine to finish the task.

"I was surrounded by my colleagues. As they watched me, they all held their breath. I drilled the first hole. When it was measured, I was so nervous that my heart beat fast." After many years, Hu Shuangqian is still very excited when he recalls the experience. "Before I started I tried the operation on similar waste materials several times. If the drilling angle was not correct, the part wouldn't fit and would be useless," he said.

It took Hu more than an hour to drill the 36 holes, each of which was immediately measured with a micrometer when it was finished. If each turned out correct, Hu would proceed with the next. It was like carving patterns on the titanium alloy part. Finally, the operation turned out 100% accurate. It was directly delivered for installation, and much time was saved.

Hu Shuangqian shares his experience with young colleagues.

"The more difficult the task, the more the sense of achievement. When the parts I make pass the tests I feel greatly relieved and not tired at all, even after working long hours." Hu Shuangqian said that it was quite fascinating to overcome challenges with his own hands.

Hu's colleague Cao Junjie, a young fitter, told us, "The factory always thinks of him when there is an urgent task, sometimes sending for him at midnight. Others seldom have such experience. As result, he does not have enough time to look after his family."

Hu Shuangqian is usually in the workshop six days a week. There is only one family photo in his house, which was taken in 2006. Before 2013, the family lived in a 30-square-meter old house for nearly 20 years. In 2013, Hu Shuangqian bought a 70-square-meter apartment on instalments in Baoshan District, Shanghai.

As expected, Hu Shuangqian's apartment is austere. It is not big but very tidy. A wedding photo of the couple and several craftworks featuring Shanghai in the 1980s are placed on a cabinet. The most eye-catching item is a commemorative model of the ARJ21 aircraft painted red and produced by the SAMF.

Although Hu Shuangqian is a highly skilled fitter, his salary is far from enough to offer his family a comfortable life in Shanghai. But over the years he has brought home many awards and certificates he has received, which are the most valuable items his family owns.

"We don't display these certificates ostentatiously at home, but sometimes we will take them out to have a look at them." At our request, Hu Shuangqian's wife Li Julan showed us her husband's honors, which are kept in a cabinet in the bedroom and seldom taken out. She was familiar with all the certificates. "This is his first award, won in 2001. It is for excellent employees. This is the National Labor Medal." Li Julan gently lifted a plastic film off the certificate, which represents the highest honor for workers. "I have protected it well," she said.

The year 2018 was the 12th anniversary of the approval of the C919 project. The new C919 aircraft will soon be ready for testing. In COMAC's factory in Pudong, Shanghai, there are four eye-catching slogans covering an entire wall on both sides of a Five-Starred Red Flag. The slogans read, "Always working hard, tackling difficulties, bearing hardships and being dedicated." For employees of COMAC, these slogans are a true description of the past and future of their careers in developing the new-generation domestic large aircraft.

Hu Shuangqian will have his 58th birthday soon. He will retire in two years, but he thinks that it is too soon.

"My job started with the 708 Project (Yun-10), and the C919 might be my last project. I'm afraid I won't be able to engage in large wide-bodied aircraft, for I'm going to retire soon. If my age permits, I want to work for another ten or 20 years and make more contribution to China's large aircrafts. This is my wish." In the final assembly factory in Pudong, Hu watched the final assembly of the C919 aircraft. He has never produced a defective product in his life, and now his eyes are full of hope.

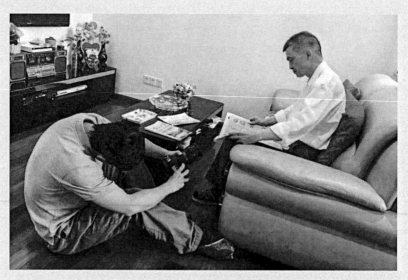

The videographer shoots Hu Shuangqian's home life.

Shanghai Artisan Hu Shuangqian

In the production of the first season of *Great National Craftsmen*, I happened to be covering the development of China's large C919 aircraft at COMAC. As the project is at the pinnacle of China's modern industry, there must be top masters working in this field.

Hu Shuangqian is one of them. Cheng Fujiang, then COMAC's publicity director, highly recommended Hu Shuangqian to us. We decided on him as the hero of our episode.

Hu Shuangqian was born in Shanghai in 1960, which makes him the same age as my father. When we got more familiar, I began to address him as Uncle Hu.

Uncle Hu is not talkative. To make the program, I had three lengthy interviews with him in the workshop, the factory and other places. Uncle Hu was quite efficient and accurate in communication. He could summarize accurately his 40-year working experience. In the program, he described the *si*-level precision in the following words: "When the bolt is placed into the hole, one can feel a little tightness, but it is not very tight, which is the correct feeling." His description was very vivid.

A major part of the interviewing was done on the lawn in the middle of the factory. We sat on the lawn, and talked casually for a long time. Behind us was the last Yun-10 aircraft made in China, as mentioned in this report. It is a monument to the flight dreams of China.

I was born in the 1980s, and I know my world and his are far apart.

But I was willing to listen to his story, and I wanted to spread it to more people.

I still remember our longest conversation. We talked from 3 p.m. till sunset. In order to ensure the unbiased color cast, our cameraman Duan Dewen adjusted the color temperature three times.

After I got to know him better, I began to define Uncle Hu as the representative of "Shanghai-style artisans." Different from the "northern artisans" who are usually brawny and forthright, Uncle Hu has a calm appearance and always pays close attention to detail. He has an inherent passion for technology, and strong professional ethics. Hu Shuangqian is a master of the game of bridge, and he likes assembling and repairing old radios in his spare time. He is warmhearted and often cuts the hair of his apprentices and workmates. He goes to work by bike. When he is excited in conversation, he cannot help uttering a happy exclamation... These details could have rendered Hu Shuangqian more vividly, but they were not adopted due to the limited program length. There were so many details that the editor Xu Xiaoran and I just didn't know which ones to pick. We wrote a memo for each detail, and discussed and deliberated repeatedly which should be kept and which rejected, as the program was less than eight minutes long, too many details would be cumbersome. Therefore, only the most important and essential parts were kept in the final version.

For *Great National Craftsmen*, we adopted a high-standard shooting method which is rare for CCTV's news channel. We used an SLR camera with three zoom lenses instead of the handy Panasonic 3100 high-definition camera. It was quite successful, being an essential factor for the core competitiveness of the program. The SLR camera offered high color saturation, which highlighted the metallic texture of the industrial production line. Our cameraman Duan Dewen was excellent in selection and application of lenses, especially the use of telephoto and micro lenses, which highlighted the image of Hu Shuangqian against the setting as well as the texture of industrial meticulous operations in the boring, milling and grinding procedures.

However, in my opinion, the essential factor behind the success of *Great National Craftsmen* was that the program focused on craftsmen who work in a down-to-earth manner, like Hu Shuangqian, and reveals the valuable qualities in them which are required in our society today.

Shanghai is undergoing rapid economic development. In this fast-paced city, Hu Shuangqian still goes to work on his old bike, just as he did decades ago.

However, he doesn't lack chances to earn good money.

As a fitter in the aviation industry, Hu Shuangqian is entitled to the much higher salary paid in all modern industrial enterprises. In the 1980s, Shanghai Aircraft Manufacturing Factory was in recession, while automobile manufacturing in Shanghai was enjoying a boom. Thriving companies sent representatives to line up in front of the troubled factory, ready to take its skilled people away. I believe that even today, few people would not be tempted in such a situation.

In my opinion, the reason Uncle Hu waited till he could fully show his capabilities after China's big aircraft industry thrived again was that he simply stuck to the most valuable thing in his mind, eschewing material gain.

After the program was broadcast, some people commented that China's aviation industry was still underdeveloped, or there wouldn't be any manual work; that developed countries had adopted fully automatic assembly lines. These comments show that the general public still has too little knowledge and too much misunderstanding of modern industry.

As I mentioned at the beginning of the program, aerospace manufacturers, even the advanced ones like Boeing and Airbus, have to spend a lot of money to keep senior craftsmen to deal with emergencies and urgent tasks on the production line. In fact, there are basically no two identical planes in the world. Even if two planes look 99% alike, one plane may have a smaller gasket, or the other may adopt a different fastener connection method. Manual adjustment exists in all aircraft.

This is certainly true of the aviation industry. Even in Lockheed Martin's Skunk Works, the most advanced in the world, artisans are indispensable for the most difficult manual operations.

This is the value of craftsmanship.

There is another thing that touched me deeply, which is not shown directly in the program: Uncle Hu had been working on the frontline of aviation manufacturing for more than 30 years. After I got to know more about him, I found that this kind of perseverance is totally different from the "muddling along" mentality common in some people.

Those with the "muddle along" mentality is content with the lowest-possible standards. They are rarely motivated to focus on their career, and they may not miss it when they lose it.

However, craftsmen like Hu Shuangqian accumulate experience in the process of making each and every product. In this process, they make themselves irreplaceable, and keep improving their skills to a higher level.

Hu Shuangqian's skill gets more consummate with each passing day. Although this cannot help him earn more money directly, it has enriched his life and career.

From the 1980s to the beginning of this century, China's aviation industry and even the whole national defense industry were at a low ebb. In some cases, the sense of responsibility and dedication in one's career could easily yield to the needs in material life. In this process, those who have stayed to promote the development of national core competitiveness are also great heroes, just like the martyrs who sacrificed for the country and the people. We should pay tribute to and learn from them.

They are peacetime heroes.

By Zhao Zhongliang, former reporter of
CCTV News Center

Hu Shuangqian and reporter Zhao Zhongliang pose in front of a C919.

MA RONG

Engraver of RMB Notes

Profile

Ma Rong is a senior artist of China Banknote Printing and Minting Corporation (CBPM), working in the Design & Engraving Division, Technology Center of CBPM. She is a fourth-generation engraver of Chinese banknotes as well as China's first female engraver of figures on banknotes. Her works include the figure of Chairman Mao Zedong on the banknotes of 10, 20, 50 and 100 *yuan* currently in circulation.

With nearly 40 years' experience, Ma Rong is qualified for independent engraving of high-standard banknote templates. She has compiled teaching materials on manual engraving and freehand-sketching. Her major works include the essential part of the design of the fifth set of RMB banknotes and design of gold and silver commemorative coins for the 2008 Beijing Olympic Games. Her scientific research has reached the advanced international level, and she has been invited to give lectures in Italy.

Ma Rong, senior artist of the Design & Engraving Division, Technology Center of China Banknote Printing and Minting Corporation (CBPM)

Preview

The 100-yuan RMB banknote currently in circulation adopts the internationally advanced recess engraving technology. The surface is made uneven for anti-counterfeiting purpose. The last step in anti-counterfeiting is the figure engraving. The figure of Mao Zedong on the banknote was carved by Ma Rong.

M a Rong has nearly 40 years of experience in recess engraving, and is China's first female engraver of figures on RMB banknotes. The following is her understanding of the difficulties in banknote engraving:

"Figure engraving is the most difficult part of banknote recess engraving. We must represent the essential features of the figure, the sense of space and the texture through dots and lines."

The difficulty of figure engraving on banknotes also lies in its particularity. "Figure engraving is the last line of defense," Ma Rong said, "It must be extremely accurate to ensure the anti-counterfeiting effect."

Whenever Ma Rong has some free time, she will instruct students in their engraving techniques. "While carving a line, we must hold the breath, grasp the knife firmly and push it forward slowly."

When asked about her impression of the hardest aspect of the process of learning engraving, Ma Rong said, "The process is irreversible."

Irreversibility means that when an engraver carves on a steel plate with a steel knife, there is only one chance to carve each line, and no possibility of "erasing" the marks and starting again. More or fewer, deeper or shallower, thicker or thinner, these traits of the marks will seriously affect the engraving. Ma Rong and her colleagues must be extremely careful with their knives. "You can only carve more or deeper, but not less. If anything goes wrong, the image will be distorted. For example, a slightly excessive strength in the fingers will make the figure's skin color too dark."

Although Ma Rong has nearly 40 years of experience, she never ceases practicing figure engraving.

It takes at least ten years to train a qualified engraver, for a lot of practice is required. Ma Rong's supervisor was strict with her apprentices. Ma Rong summed up her supervisor's secret to success as "practicing again and again."

Zhao Yayun, the third-generation recess engraving master of China, is Ma Rong's first supervisor.

Although Ma Rong mastered the craft many years ago and has now achieved great success, she is always deeply indebted to Zhao Yayun for her training. Every year, she will visit Master Zhao with her latest works.

This time, Ma Rong brought her work titled *Bronze Ware* to Master Zhao. Master Zhao was very happy to see Ma Rong. She held Ma Rong's hand intimately

Ma Rong at work

as if Ma Rong was her daughter. She looked at Ma Rong's work closely as usual, observing each line under a magnifying glass. She was very proud of Ma Rong. "At the beginning, the apprentices could not focus on the work. I would help them to calm down by getting them a glass of water and asking them to observe it. I skipped this step for Ma Rong, because she was patient enough. She had only one idea – that she must master her job."

Ma Rong adores Master Zhao. "She was the first female engraver in our country. I admire her very much. When I started learning this trade, I was determined to become an excellent engraver like her."

Master Zhao was quite strict with her students, but Ma Rong set higher standards on herself on this basis. "If I made a mistake, I would blame myself. If I was distracted or couldn't calm down, I would do it all over again, forcing myself to totally focus on the engraving. I was so concentrated that I was unaware of anything else. This is the best state to be in."

In order to achieve this "best state," Ma Rong practiced numerous times. In the end, she felt that her hand, the knife and the steel plat had become one: "I felt that the steel plate had become softened."

The hard steel plate becomes soft under Ma Rong's hands, which is the most desirable state for Ma Rong' students.

Niu Kai, a student of Ma Rong, once asked her: "What is the best state like?"

"When you are carving, your hand should be so relaxed that you can move your fingers deftly. Only in this way can you produce vivid lines," said Ma Rong.

As the first female engraver of figures on RMB banknotes, Ma Rong knows well the functions of the figures: "Banknotes with figures are more eye-catching. The figures are usually public figures or political leaders of a country. The banknotes are made in such a way that people can instantly tell whether one is real or fake."

It is the dream of all engravers to apply their skills to banknotes.

When a new version of renminbi banknotes was to be released in 1997, Ma Rong and other banknote recess engraving artisans competed for the opportunity of engraving Chairman Mao Zedong's figure on them.

Unfortunately, Ma Rong's entry was not selected. "At that time, I was not quite experienced. When I was carving, my knife turned a little bit outward, and an eye of the figure got a little too big. I lost."

The journalist asked her: "Were you sad at your failure?"

"I realized that there would be more chances," Ma Rong said calmly.

Ma Rong never ceases to work at improving her skills. She said, "The images must be accurate, refined and imbued with the engraver's personal traits. An engraver must try his or her best to produce works people like."

A few years later, Ma Rong had another opportunity. This time, the artisans competed for the chance of engraving Chairman Mao's image on the 20-yuan RMB banknote. "At that time, senior masters had retired one after another. As we were much younger, that is, 20 to 30 years younger, we felt great pressure. If we could not do it well, the quality of banknote engraving would fall."

"What will happen if the quality falls?"

"It is an anti-fake measure. If the quality drops, it will affect the security of the currency."

Ma Rong had many excellent ideas, but she couldn't realize all of them on one steel plate, for only one figure could be carved on one plate. Therefore, Ma Rong made a bold decision to carve two steel plates in the competition, which meant double time and efforts.

Ma Rong gives instructions to a student.

Ma Rong shows the reporter how to judge the authenticity of a RMB banknote by feel.

"When I finished 80% of both plates, time was running out, so I had to give up one plate. I was satisfied with one version in which the eyes of the figure were more vivid, which endowed the figure with a more distinctive outlook. Now the figure is used on the 50- , 20- and 10-yuan notes."

Figure engraving is quite demanding, so Ma Rong keeps reminding her students: "You must be extremely careful."

Rigorous craftsmanship has its unique use in banknotes. Ma Rong has profound understanding in this regard: "It is an anti-fake measure, so an extra dot or the lack of a single line indicates a fake. Each is unique, and an engraver might not necessarily carve every plate exactly the same."

On learning that Ma Rong's plate had been adopted for the 20-yuan banknote, her supervisor Zhao Yayun was very surprised. "I didn't believe it had been done by Ma Rong," said Zhao Yayun, "I thought it had been done by some retired senior master. For this achievement, she must have practiced a lot after I retired. Carving Chairman Mao's figure for banknotes was not easy job."

For Ma Rong, the most difficult part of banknote recess engraving is not engraving on a steel plate, but transformation. Internationally, with the upgrading in technology and equipment, many items of equipment for recess engraving are no longer available, and recess engraving must have its transition, too.

It was a transition from traditional engraving to digital engraving.

"At first, we felt that manual carving is an art, but carving with the aid of computers would not be the same. However, when I realized that we were to shoulder the responsibility of RMB banknote plate engraving, and that we had to attain a higher level with the aid of new technology, I also realized that this transition would be a tough process of thorough change."

Ma Rong began by learning basic computer skills. She bought a computer with one month's salary, and learned at home. After work, she practiced the basic skills of computer operation that she had learned from her younger colleagues during the day. She took notes carefully: "All tools and documents are well sorted out and can be easily accessed."

Speaking of the difference between digital engraving and manual engraving, Ma Rong said: "The stylus has been replaced by the mouse."

The figure on the newly issued 100-yuan banknote was created by Ma Rong with digital aid. Ma Rong transplanted the entire recess engraving craftsmanship into the computer, and summed up her practical knowledge for digital engraving, realizing the perfect transition from traditional craftsmanship to modern technology.

Ma Rong knows well how to judge the authenticity of banknotes. "You can tell by feeling the texture of the paper along the direction of the lines. Besides, in counterfeit notes, there is more color change."

Ma Rong's husband Kong Weiyun is a banknote recess engraving artisan, too. He has great affection for his job. He said, "Recess engraving has a history of more than 100 years in China. It was introduced to China from America. The Chinese artisans pass the craftsmanship from generation to generation, and have created many classic works in banknote plate engraving."

Ma Rong and Kong Weiyun used to be classmates, and they are colleagues now. Kong Weiyun has a more refined and exquisite style of manual engraving. The couple often draw inspiration from each other in their creations.

They have collaborated in creating a masterpiece of banknote recess engraving.

Ma Rong showed the reporter a one-yuan note in circulation: "The figure of Chairman Mao on the front was carved by me, and the scene of the West Lake on the back was completed by my husband."

As an accomplished RMB banknote recess engraving artisan, Ma Rong was invited to give lectures for two months at the International Sculptors' Academy in Italy. Kong Weiyun helped her to prepare the teaching materials. Kong said, "Take this *Riverside Scene at the Qingming Festival*. It was jointly created by some of our artisans. It can demonstrate the engraving craftsmanship as well as the traditional culture and art of China."

Today, the number of recess engraving artisans is dwindling throughout the world. Therefore, an international academy for banknote engravers was set up in Italy. Artisans from all over the world can study here. Top masters are invited to give lectures, including Ma Rong, who shared her craftsmanship and experience with her peers.

Kong Weiyun knows Ma Rong very well. "I think it is our responsibility to introduce Chinese engraving craftsmanship and traditional culture to the whole world via the RMB banknotes," he says.

Ma Rong has been engaged in RMB banknote figure engraving for four decades, and she has developed an unusual affection for the currency. "There are many convenient means of payment now, but I still prefer payment in cash. Others might be more concerned with the face value of the notes, but I pay attention to the banknotes themselves as well, for they carry part of Chinese culture. I cherish and value the RMB banknotes very much."

At present, there are only a dozen artisans qualified for banknote engraving in China. Ma Rong belongs to the fourth generation of these artisans, and she is now training the fifth generation.

Ma Rong communicates with her steel plate through each and every dot and line. In her mind, every stroke of the stylus is irreversible, just as her course of life is. Over 40 years, Ma Rong has been pursuing her lifelong career with a stylus and her great tenacity of a craftsman. She has carved herself an excellent course of life.

Great National Craftsmen
in the New Era

China's recess engraving technology on banknotes is currently the world's top anti-fake technology. Apart from ensuring the maximum anti-fake function, the artisans also pursue perfect artistic effects. For decades, the artisans have been doing their work diligently, and guard the safety of RMB banknotes from the source.

Two years ago, we had a very quiet time in the design and engraving room of the technical center of China Banknote Printing and Minting Corp. After the program was broadcast, I wrote down the following internal monologue on my WeChat album:

"She is linked to each of our RMB notes –

Each stroke of her stylus is irreversible,

Just as her life course

In silent perseverance

Of superb craftsmanship."

In the *Great National Craftsmen* series, I participated in the coverage of the first five seasons of news programs and the first episode of documentaries. Among the programs, this episode about Ma Rong is one of my favorites. The artisans that I have interviewed have a common point, which can be summarized as "a naturally pure feeling." In the words of Cui Wenhua, the chief contributor to the documentary *Great National Craftsmen*: "The greatest craftsmen serve the country and the people wholeheartedly with their superb skills. By improving their skills and elevating their minds, they better serve others." The artisans in the programs are all dedicated to their careers. The program about Ma Rong was an episode of the third season of the *Great National Craftsmen* series. If we were still to try to impress the audience with the previous approaches, it would be somewhat boring. Therefore, we had to innovate in the new episode in order to retain the brilliant success of the program. In order to make a breakthrough, Xiao

Zhensheng, director of the Economic News Department of the CCTV News Center, called a meeting of all the producers of the department to solicit innovative ideas for the third season of *Great National Craftsmen.* Song Jianchun, the producer of the agricultural economy group, once worked with Director Jiang Qiudi in making the program *Common Concern,* the first public welfare program made in China by the Publicity Department of the CPC Central Committee. Jiang is good at reporting characters. At the meeting, he offered a suggestion: "Can we use the first-person perspective in the narration?" We adopted his proposal in the creation of the third season. I was given the task of making a sample film when I was on a business trip. The schedule was tight. I was lucky in having Jiang Jianchun as my producer. As I became familiar with him, I often asked him for advice to learn how to use first-person perspective in the program.

It was in Ma Rong's studio that I met her for the first time. It was very quiet. A journalist from Workers' Daily was taking pictures of her, so I first talked to Mr. Kong Weiyun, Ma Rong's husband. According to my experience in the previous two seasons of *Great National Craftsmen,* the secret to a successful figure report was to talk about the figure and get as much related information as possible. Kong Weiyun was very patient, and I learned from him about the history and process of banknote recess engraving. But the information I got was far from enough to make a feature report on Ma Rong. I told Ma Rong and Kong Weiyun that I would stay with them while they were working, having meals, and interview them at home.

This episode has a serene and aesthetic style, which is inseparable from the special process of recess engraving. Ma Rong's workplace has a calm atmosphere. Everyone walked and spoke gently. The technicians were carving, but they seemed immobile. A few days later, I got used to the atmosphere, and even became gentler myself. I wondered if the word serenity could become the key word for this episode. Shao Chen, the camera man, and I worked hard in this direction. To highlight this serene atmosphere, we shot in off-work hours when others had left the company.

Several days later, we collected video materials about Ma Rong: how she sharpened her stylus, how she carved on a steel plate, how she instructed her apprentices in carving techniques, etc. Yet I felt that these pieces were still only fragments and were not enough to establish a vivid image of a great artisan. We entered a bottleneck which was difficult to get through. Therefore, we switched our shooting rhythm and scene. I remember clearly that we visited Ma Rong's first supervisor Zhao Yayun on a Thursday, the deadline for our task. Ma Rong and Kong Weiyun took us to Zhao Yayun's home in a dormitory building of Beijing Printing Plant.

Zhao Yayun, almost 90 years old, is a third-generation recess engraving master of China. On seeing Ma Rong, the old lady held her hands for long time, as if Ma Rong were her daughter she had not seen for a long time. She repeated in sonorous voice that "it is worth my life that I had Ma Rong as my apprentice." The lively opening scene was a breakthrough for our interview, made previously. Zhao Yayun talked about how she had taught Ma Rong, how she passed on skills to her and her impression of Ma Rong. We got a much more concrete idea from this about how Ma Rong learned figure engraving, and her image as a great artisan became more vivid.

The visit to Zhao Yayun opened a window for our interview. In the old lady's passionate eyes and words, we felt the profound affection of the artisans of figure engraving. In the evening, I sorted out the interview materials, drafted a report and sent it to Director Jiang Qiudi. She replied, "Your report is smooth and coherent. The essence of Ma Rong's character is her serenity. Please proceed by contacting the machine room." The next morning, I met Hu Haichao, the video editor of my program. To this day, I still remember and appreciate his help. I gave him a huge amount of materials, totaling over 1000 GB, and telling him that there would be more; I was going to shoot more on the same day. Haichao agreed to my plan, and immediately started to upload my materials. When I returned from Beijing Banknote Printing Factory in the evening, he had already built the program's framework according to my script. While he was uploading that huge amount of video materials and building the framework, he understood the idea that I wanted to express in the program. Haichao showed me a few sample pieces he had made, and discussed with me how to design the beginning and end of the program. The episode was done from the first-person perspective. There was a section of more than 10 minutes that was too stagnant and lacked rhythm. We added some scenes of the shooting site to adjust the rhythm. We also added some plain-style video materials with music to show the details of Ma Rong's work. On Sunday afternoon, Director Jiang Qiudi came to check the program, and said that it basically reached her expectation.

After modification and improvement, the episode was to be broadcast as scheduled, but something unexpected happened. On that day, only three minutes were allowed to report the episode on *CCTV News*. However, Sun Yusheng, the leader on duty that day, watched the program and thought that some excellent content in the complete version should be added in the report on *CCTV news*. We had only ten minutes. As experienced editors, we were good at deleting content. As for adding more content, I had no time to think about whether I could catch up with the broadcast, but had to do it in the shortest-possible time. Fortunately, we succeeded after a few minutes past seven, when our program was sent to the broadcast room. It was not a refined piece of work, but thank goodness we didn't miss the broadcast!

Over the years, the *Great National Craftsmen* series has exerted a great influence. The programs advocate a positive social ethos and unite the thinking and strength of society. Our era calls for craftsmen and provides the soil for the growth of craftsmen. Ma Rong is totally devoted to her career. Her diligence and perseverance make her a great master. The production team of *Great National Craftsmen* also deserves the title of master. Take Yue Qun, the leading producer, for example. I have cooperated with her many times, and benefited a lot from the experience. In the production of the third season of the *Great National Craftsmen*, Yue Qun's husband fell ill suddenly, and she had to take care of him. Despite this, Yue Qun was concerned about my progress, and helped me as much as possible. In producing the programs, my colleagues and I often worked long hours in the editing room. We encouraged each other, and came up with many innovations.

In May 2017, after the sample of the fifth season of *Great National Craftsmen* was completed, I wrote the following words in my WeChat, marking the end of the work of this stage and a new beginning:

Great masters' infinite dedication in tangible crafts

Can be an ethos influencing and uniting society

In the past two years, five seasons of *Great National Craftsmen* have been broadcast. I admire the great masters in the programs, and would like to thank all my colleagues for their dedication and hard work.

Let's work hard together and enjoy it!

By WANG KAIBO, reporter of
China Media Group & CCTV News Center

Ma Rong and Kong Weiyun with the interview team

ZHANG DONGWEI

Welding Steel Plates on LNG Ship

Profile

Zhang Dongwei, born in Shanghai in December 1981, and holder of a junior college degree, is a senior technician and leader of the 2nd electric welding squad, enclosure system workshop of the 2nd General Assembly Section of Hudong-Zhonghua Shipbuilding (Group) Co., Ltd. He mainly engages in carbon-dioxide and argon-arc welding of the enclosure system of LNG (liquefied natural gas) ships.

Zhang Dongwei assiduously studies shipbuilding technology with a craftsman's spirit. He has become a backbone worker in building the company's top-end 45,000-ton container ro-ro LNG ships, which are of the world's most advanced level and most difficult to build. He dedicates himself to the construction of national marine equipment.

Zhang Dongwei, senior technician of Hudong-Zhonghua Shipbuilding (Group) Co., Ltd.

Preview

LNG ships, known as "super freezers on the sea," carry liquefied natural gas for long distances at extremely low temperatures – below –163°C. In civil shipbuilding throughout the world, the construction of LNG ships is as difficult as that of an aircraft carrier. Currently, only a few countries such as the US can build LNG ships. China didn't have any technician who had mastered the welding technology for LNG ships until 2005, Zhang Dongwei being one of the 16 technicians in the first batch.

Zhang fastened his belt, and put on his sheepskin gloves and protective mask. As the familiar droning of welding sounded, a new day's work began.

Zhang said: "I enjoy the arc light and the sound in welding. It is like music, and makes me comfortable."

Zhang Dongwei is welding the inner tank, or the core part of China's ninth LNG vessel. The welders have to connect pieces of invar steel plates like sewing pieces of cloth. The invar steel is quite thin, the thinnest part being only 0.7 mm and resembling a piece of craft paper. Behind the steel plate are wooden boxes which might catch fire from any slightly excessive welding. No wonder some say that welding seams is like playing with fire on wood.

An hour passed. Behind Zhang Dongwei's mask which reflected the arc light, sweat was running down his sideburns. In invar steel welding, the biggest challenge is to keep a stable mental state, which is not an easy job, but Zhang Dongwei can always manage to be calm when he is welding.

It takes four seconds to walk 3.5 m, but it takes Zhang Dongwei five hours to weld a 3.5 m-long seam. In the five hours, Zhang Dongwei remains absolutely still, only his hands operate steadily and gently as if stroking feathers. He is indeed an excellent technician.

Zhang said: "I can recognize the seams I weld. They are very neat, like fish scales. I like to create an embroidery effect in my welding, with all the 'stitches' distributed evenly."

According to Zhang Dongwei's colleagues, "He often works for three or even four hours without a rest."

Welding for long hours nonstop challenges a welder's basic skills. One must be able to squat for long hours and keep one's hands steady. Zhang Dongwei said that if one fails to do so, the continuity of the welding will be disrupted, which will affect both the appearance and quality of welded seams. Therefore, a welder should practice the basic skill of keeping a firm horse-riding stance, just like practicing kongfu.

Invar steel manual welding is the most demanding welding technique in the world. Zhang Dongwei's first supervisor Qin Yi was the first welder in China to master this skill. When Zhang Dongwei was in his twenties, Qin Yi demonstrated his single-side welding and two-side forming skill to him for the first time, which was a big eye-opener for the young man. In the two techniques, while the operator

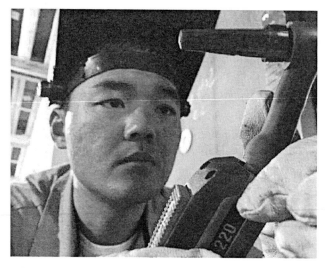

It's quite common for Zhang Dongwei to work three or four hours nonstop.

is welding on one side of a plate, a neatly welded seam is formed simultaneously on the other side. The molten iron was perfectly obedient under Qin Yi's hands. Zhang Dongwei was deeply impressed by his supervisor's superb skills, and he was determined to acquire them.

Invar welding is harmful for the health. The intense arc light is bad for the eyes, and the hot metal droplets produced in welding may hurt the operator. While learning the trade, many apprentices would watch for an hour or two and then leave, but Zhang Dongwei always stayed by his supervisor.

Zhang said: "If I stood far away, I would never master the skills. If my supervisor welded for seven hours, I would watch for seven hours. Although the repeated operation seemed simple, it took one or two years to learn."

In order to grasp the welding of an LNG vessel as soon as possible, Zhang Dongwei often observed his supervisor at work for long hours. He tried to learn everything by heart without missing the smallest detail. As the saying goes: The supervisor teaches the trade, but the apprentice's skill is self-made. In order to be as skilled as his supervisor, Zhang practiced a lot to obtain the correct feeling. Sometimes while having meals with his family at home, he would move his chopsticks in the air for practice. His daughters thought it was funny, and imitated him.

LNG vessels are mainly used for loading and long-distance transportation of liquefied natural gas at −163°C. It is internationally recognized as a "three-high" vessel – with high technology, high challenge and high added value – and is known as the "pearl in the crown of the shipbuilding industry." In the past, only a few developed countries such as America, Japan and South Korea and some countries in Europe could build LNG ships. As there was no precedent to follow at home, plus the technology blockade implemented by other countries, building LNG vessels used to be an ordeal for Hudong-Zhonghua Shipbuilding (Group) Co.,Ltd., which could only grope ahead step by step.

At first, foreign countries didn't believe that China could master the technology. To qualify for invar welding for LNG vessels, welders had to pass a strict examination of the international patent company of GTT. Even if they obtained the qualification certificate, evaluations were carried out on a monthly basis to ensure they remained qualified for the job. Among Qin Yi's first group of apprentices, Zhang Dongwei was the first to pass the GTT exam. Foreign examiners gave Zhang the thumbs-up and wrote a big "OK" on the steel plate he had welded for the exam.

Zhang Dongwei said, "When I was an apprentice, I was determined to surpass my supervisor. Now I have apprentices, too, and I hope that they can also surpass me. If everyone can make their own improvements, our nation will not lack technicians in this trade."

The LNG building team of Hudong-Zhonghua Shipbuilding Group once set a record of zero leakage in a large tank in a third-party welding test. It is amazing that there was not a single leakage point in the welding seams with a total length of 35 km. Apart from superb skills, it requires extraordinary patience and concentration. Therefore, if one can't keep calm and patient in welding, one cannot complete the task with perfect quality and quantity.

The ultra-low-temperature resistant invar steel is as thin as paper and quite easy to become rusty. The thinnest place can rust and be totally damaged by one touch in the first 24 hours, so there should be no sweat or fingerprint in the welding. The welders must be steady physically and psychologically, for any fluctuation in mood may affect the quality of the welding. In order to achieve such a stable mental state, Zhang Dongwei often goes fishing in his spare time.

Zhang Dongwei cultivates a tranquil mind by fishing.

Zhang said, "My job might be boring to some people. It requires much patience. I like fishing, because it is somehow similar to my job. I can keep watching the fishing float for long time, probably for eight hours."

Zhang admitted that, compared with other trades, the shipbuilding industry is not attractive; on the contrary, it is very hard. He faced a lot of temptations. Once, Zhang went to a welding exposition, and took part in a welding match there. After the match, several factories wanted Zhang Dongwei to work for them, offering to double his salary and other benefits.

However, Zhang Dongwei refused all such offers. Since graduating from Hudong-Zhonghua Vocational School, he has been working for Hudong-Zhonghua Shipbuilding. He became specialized in the building of LNG ships. For more than ten years, Zhang Dongwei has worked diligently and quietly, contributing to the company as well as realizing his own value. Apart from enhancing his own skills, in order to cultivate more capable technicians, he also passes his knowledge and experience to his colleagues without reservation. In the ten years from 2005 to 2015, thanks to Zhang Dongwei's instructions and guidance,

Zhang Dongwei shares his experience with his colleagues.

over 40 staff members obtained related certificates, including the top-level invar welding G certificate, SP3/SP4/SP7 hand welding certificate, and MO1-MO8 argon arc automatic welding certificate, covering all welding types and meeting all demands of the containment system of LNG ships. Zhang has trained more than 30 composite-type invar welders, two of whom are now team leaders, while the rest are all technical backbones of the workshop.

As Zhang Dongwei is very busy, he does not have much time for his family. He often only gets home once a week, and when the ship delivery time is tight, he will work extra hours at weekends. When his daughters miss him, they will draw pictures of ships. In the girls' mind, their father is a "super hero" building big ships.

Zhang Dongwei: "The ship Daddy builds is as high as a ten-storied building."

His daughter: "Wow, that's really big!"

"Daddy is as small as an ant on the ship."

"How about me?"

"You would be like a small ant."

Zhang Dongwei said he had no grand dreams when he was a child. His goal was not to go to a good high school and then enter a famous university, but to have a stable job and live an ordinary life. He didn't know how hard it was to be a welder and how much harder it was to be a good one until he chose the trade. It is tough to wear a heavy smock and deal with scorching steel plates, smoke and dust all day long. In summer, other people feel hot even when they are wearing thin clothes, but welders must wear thick smocks. Zhang Dongwei's career experience is both simple and extraordinary. He has been making progress with patience and tenacity far beyond those of his peers. He never flinches in the face of difficulties and challenges.

Zhang Dongwei said: "Whatever challenges I meet, I have never thought of giving up, not once."

Up to now, Zhang Dongwei has participated in the building of all China's 12 LNG vessels.

"When I see an LNG vessel launched on its maiden voyage, I feel very proud. Unlike tangible things, for example, documents in computers that can readily be printed out, skills can only be obtained through learning and practicing over and over again. Only those genuinely interested in it will learn it by heart. It is an inner pursuit," Zhang Dongwei said.

As more domestic LNG vessels sail around the world, Zhang Dongwei and other invar welders' thoughts will travel farther along with the vessels they have produced.

The videographer shoots the work scene of Zhang Dongwei.

Cultivation of Mind First,
Skill Next

Three years ago, with the debut of the *Great National Craftsmen* series on CCTV, eight great supervisors came quickly into the public view. Calling for great artisans and carrying forward their spirit have become a lasting hot social topic. Among the great artisans, Zhang Dongwei is the only one born in the 1980s. He has drawn more attention probably partially because of his youth.

After the program was broadcast, Zhang Dongwei became better known, and received more honorary titles, such as "National model worker of professional ethics construction," "National May Day labor medal" and "China quality award nominee." Did these honors come too early for him? Could this low-profile young man handle the honors that came almost overnight? Would he be flattered? Or could he stay undisturbed and focus on his skill?

At that time, I was actually a little worried about him.

But three years passed, and with them my worries and doubts gradually disappeared.

Every time I contacted Zhang Dongwei, he was as modest and gentle as before, always addressing me politely. When I asked about his current status, he said simply: "It's almost the same as before – working and training my apprentices." Zhang is still the team leader of his workshop and technical leader of invar welding. I once joked, "Dongwei, there is an entry about you in Baidu Baike (a Chinese website similar to Wikipedia), so you are a celebrity now." He just laughed it off, and would say no more about it. He was so calm that I couldn't tell if he was at all excited about or proud of his fame.

Sometimes I wonder how such people become great national craftsmen. Is it because they can endure loneliness and resist temptation? Craftsmanship and expertise are certainly not enough. There must be cultivation of the mind first and then development of skills. Only those with a serene mind can deal with issues with great composure. In Zhang Dongwei there is a detachment and aloofness that is beyond his age.

However, three years ago, when the subject of my episode of *Great National Craftsmen* was not yet determined, I was not sure if he was suitable to be the subject.

On April 17, 2015, I was told to shoot an episode of *Great National Craftsmen,* a news program on the subjects. The choice of interviewee was crucial to the success of the episode. We held a meeting at which some colleagues named their interviewees. They were basically experienced old supervisors with calloused hands, having seen much of life and having a lot of touching stories. They fitted the concept of "Great National Craftsmen" in my mind. Although I hadn't finally decided on my interviewee, my anticipation was similar to that of my colleagues.

I met Zhang Dongwei at Hudong-Zhonghua Shipbuilding (Group) Co., Ltd. for the first time. The tall young man I met completely overthrew my stereotype of experienced supervisors. As Zhang was only just over 30, would his story and experience be sufficient for the program? I was a little worried.

At Hudong-Zhonghua Shipbuilding Group, I learned that China had started invar welding about 10 years before, so the post-1980s generation was indeed "old supervisors" in this trade. Moreover, invar manual welding was a top welding technology in the world. In the management's eyes, Zhang Dongwei was definitely a suitable candidate for my program. Half convinced and half doubting, my cameraman Li Ziguo and I followed Zhang Dongwei closely for four days. Zhang and I were of the same age, so we had a lot in common. A few days later, I knew more about Zhang and his invar welding.

Zhang Dongwei is working in the ship's cabin. For safety, everyone entering the cabin is required to wear special apparel from head to toe, including a pair of steel shoes which weigh several kilograms. It is very difficult to walk on land in the steel shoes, let alone walking in them inside the huge 10-story cabin, where elevators are used to transport goods only, not people. We followed Zhang Dongwei climbing up and down in steel shoes. My heels were rubbed raw in less than a day. When the first day was over, I could barely move my legs. However, this was the everyday experience of the invar welders.

Zhang Dongwei and his colleagues seem no different from other ordinary people who commute in the busy city every day, but they are also extraordinary. In a world full of temptations and distractions, it is rare and

Zhang Dongwei (right), Zhang Dongwei's colleague and Guo Wei (middle), the reporter

difficult to choose to ignore these temptations and concentrate on work. However, these young men do it. They inherit this "legacy" from their supervisors with great sincerity and reverence. Once they make the choice, they engage in their career with great dedication and commitment, and will never give up.

I am glad that I chose Zhang Dongwei, and presented the young craftsman on the screen. He is a "Great National Craftsman" of calmness, independence, perseverance and dedication.

Choose a career, and be dedicated to it throughout life. This "craftsman spirit" is the ethos of the era, and exerts a profound social influence.

By GUO WEI, former reporter of
CCTV News Center

ZHOU PINGHONG

Leading Endoscopy Specialist

Profile

Zhou Pinghong, director of Endoscopy Center, Zhongshan Hospital, is a world top-level endoscopy expert. Having broken through the forbidden zone of endoscopic surgery, he is revered as a mentor by his European counterparts and almost as a divinity by his Indian peers. He and his team dominate the stage of the world's digestive tract surgeries. A surgical procedure named after him is known to the whole world. He has opened the world's fourth channel leading to life with his expertise and benevolence. A Greek painter has painted a portrait of him, saying that she liked his eyes because there was love in them.

Zhou Pinghong, director of Endoscopy Center, Zhongshan Hospital

Preview

This is a story about a doctor of superb expertise and benevolence.

In ancient China, doctors were defined in the category of craftsmen, for example, in *Huangdi Neijing (Inner Canon of Huangdi,* a seminal text of ancient Chinese medicine). But ever since ancient times, doctors with superb skills and benevolence have always been considered as noble figures who deserve great admiration.

Zhou Pinghong was born in a village in south China. He well remembers his childhood – the fragrance of rice and the hard farming work. He knew well the hardships of his fellow villagers. Since he became an excellent doctor, he has become all the more concerned with his patients. Underlying his expertise is his dedication to career, and benevolence toward his patients in particular.

Professor Zhou Pinghong works almost nonstop every day. He shares his experience with his peers without any reservation. He travels throughout China to cure patients, some of whom live in remote mountains in extreme poverty. No wonder the Greek painter said that there was love in his eyes!

From July 8 to 10, 2016, the 3rd Athens International Symposium – Gastrointestinal Cancer was held in Athens, Greece. More than 400 professionals from 25 countries and many top experts in the digestive endoscopy field all over the world attended the conference.

The conference is renowned as the "Olympic" of the endoscopy field.

In order to promote the development of minimally invasive endoscopic surgery, one important item on the agenda of the conference is to have endoscopic surgery experts conduct live operation demonstrations, after which the participants give their comments on the operations.

The first operation, also the most difficult one, is usually performed by top-level international experts. That year the organizing committee gave the opportunity to Zhou Pinghong from China.

At that moment, Zhou was waiting in his surgical gown at a Greek hospital several kilometers away from the conference venue. Surrounding him were Greek doctors and nurses who watched him in admiration. Zhou Pinghong was as revered as Hussein Bolt at the Olympic Games.

The door to the operating room opened. Dr. Zhou walked out and down the corridor. Four assistants were waiting for him at the end of the corridor, some of them world-famous endoscopy experts.

Zhou Pinghong was to operate on a Greek patient with achalasia, a condition in which food could not pass through the esophagus because of incomplete relaxation of the lower esophageal sphincter.

Before the application of endoscopic surgery, the disease could only be addressed via major surgery. Today, thanks to endoscopic technology, minimally invasive surgery can be applied via natural channels in the patient's body.

However, a surgeon cannot "see" with his naked eyes the inside of a patient's body. Different from traditional operations, in endoscopic operations, the surgeon is not operating directly on the patient. Actually, the surgeon operates at the endoscope operation end, which is 1.2 m away from the patient. With the help of the camera at the end of the endoscope, he follows images transmitted from the camera to move the endoscopic scalpel in the patient's body.

The endoscopic scalpel entered into the patient's mouth slowly, went down the esophagus, and reached the cardia.

A live broadcast of surgery at the conference

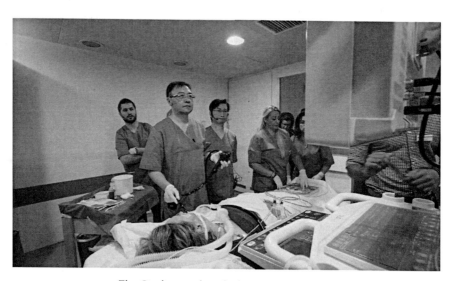

Zhou Pinghong conducts the demonstration operation.

Zhou Pinghong's surgery video, including pictures of how the endoscopic scalpel was moving inside the patient's body, was transmitted in real time from the operating room to the conference site several kilometers away.

The conference room was in utter silence.

The regular conference offers valuable learning opportunities on digestive tract surgeries, where surgical operations are demonstrated in an open way without any reservation.

The nidus is at the back of the cardia. Zhou had to be extremely careful when moving the endoscope in the esophagus to completely cover the nidus.

The assistants in the operating room and the experts at the conference all knew how difficult this surgery was.

However, Zhou Pinghong always wore a smile throughout the process. He even interacted in fluent English with experts in the conference room, telling them the key points of each step and the next steps.

In fact there were many uncertainties in the course of the surgery. Everything depended on Zhou Pinghong's judgment at every moment, and any slipup would mean the failure of the operation.

When Zhou Pinghong was cutting the nidus accurately with his scalpel, the audience were in rapt attention.

Present at the conference was Zhou Pinghong's tutor, Professor Yao Liqing from the Endoscopy Center of Zhongshan Hospital in Shanghai. Yao Liqing was confident in his student, and believed that Zhou had long surpassed himself.

The operation took only 30 minutes, which was a rare speed. An expert from Germany marveled at Zhou Dongping's superb skills, and said that it was a blessing for the whole world to learn from such a master.

This is a boon from China to the whole world.

Outside the conference room, a Greek painter and wife of the president of the conference sent Zhou Pinghong a gift: a portrait of himself. When asked what part of the painting she liked most, she smiled and said it was his eyes, which were full of love.

This was the impression Zhou Pinghong and other Chinese doctors give to the world. They pass on not only their great skills, but also, more importantly, love.

Back home from Greece, Zhou was energetic as usual. He seldom gets jet lag.

Early in the morning, Dr. Zhou went to the inpatient department, followed by his colleagues from the Endoscopy Center, intern doctors and peers from abroad.

They all followed Zhou closely, for this was their best opportunity to learn.

He stopped at a teenage girl with a pale face. She was an achalasia patient and had had surgery two day previously. The girl had not yet recovered from it because of her weak constitution.

The girl's operation had been performed by Zhou Pinghong. Now her parents were very excited to see him. They told him eagerly that the girl could take some liquid food today, the most she had eaten in a month.

Professor Zhou said with a smile: "It is good that your daughter can eat. Take it easy. she will eat normally in a few days."

This is the most gratifying answer to a patient's family.

At the Endoscopy Center of Zhongshan Hospital Affiliated to Fudan University, Shanghai, where Zhou Pinghong works, more than half of the most difficult POEM (Peroral esophageal myotomy) surgeries in the world are performed. Currently, POEM is the best treatment for achalasia.

In the international digestive tract treatment sequence, POEM is also called Zhou's surgery, as it is named after Zhou Pinghong. It is an honor from the world's digestive tract therapists for the Chinese doctor.

In endoscopic surgery, a special tubular endoscope about 1.2 m long penetrates the patient's body, and performs the surgery precisely. For patients, the biggest

Zhou Pinghong shakes hands with a patient on his ward round.

benefit is that only a small incision is made in the digestive tract, a natural cavity of the body, which eliminates the pain of thoracotomy, and reduces the risk of surgery as well as the cost to patients.

Zhou Pinghong was to perform a POEM surgery on a patient with a tumor near his cardia. According to the traditional treatment, a thoracotomy had to be conducted to remove the tumor. Now Zhou Pinghong was to remove the tumor via minimally invasive endoscopic surgery.

As endoscopic surgery is performed inside the patient's body, it doesn't require a rigorous sterile environment as in traditional surgical operations. Therefore, when other doctors, including foreign experts who come to study treatment of digestive tract diseases, learn that Zhou Pinghong is to perform surgery, they will come to watch and learn on the spot, in which case the Endoscopy Center is usually very crowded.

Zhou Pinghong is quite used to this. He even jokes with others in such an environment. His operations, which combine theory and practice, are the best opportunities for sharing his skill with his colleagues and peers.

Zhou Pinghong's operations are the most direct teaching process. In his opinion, only through sharing can he discover more problems, think more and innovate more scientifically.

The surgery is performed between the superficial mucous layer and the deeper muscular layer of the esophageal wall. The thickest part of the esophageal wall is only 0.4 cm, so operating in such a narrow space might cause greater damage to the patient's esophagus. Therefore, Zhou has invented a new approach by building an invisible cavity through the interlayer of the patient's esophagus wall.

A cavity is made between the esophagus and the stomach wall, in which Zhou will perform his endoscopic surgery.

Making a cavity in the esophagus wall is Zhou Pinghong's original creation. The first step is to separate the otherwise closed submucosa of the esophagus by injecting normal saline solution into the interlayer. After the injection, a cavity that cannot be seen by the naked eye is formed. Then a scalpel with a diameter of only 3 mm can easily enter this narrow cavity to perform minimally invasive surgery.

Zhou Pinghong explains each step-in detail to the colleagues and peers surrounding him. However, this doesn't mean he might be absent-minded in his surgery.

Zhou fixes his eyes on the screen without blinking. Under his control, the endoscope enters into the patient's body accurately.

It only takes Zhou 20 minutes to complete the operation. At present, even in Japan, which stands at the forefront in endoscopic surgery in the world, a simple POEM operation takes at least an hour. Zhou Pinghong's extraordinary skills have attracted many of the world's top experts.

On July 3, 2016, invited by Jiangsu Provincial People's Hospital, Zhou Pinghong went to perform an endoscopic surgery operation.

Among the doctors watching was a Japanese endoscopic-surgery expert named Oura. Like the other doctors, he took out his mobile phone to film the operation. Zhou Pinghong turned and said to him with a smile: "I will collect patent fees for this."

Oura was surprised. But Zhou said smiling: "Just kidding! I will continue with my operation, and you continue with your shooting."

Zhou Pinghong completed the operation in half an hour. It was amazing.

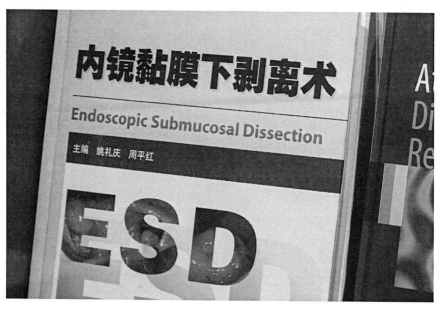

A medical book co-authored by Zhou Pinghong and his tutor Yao Liqing

Ten years ago, minimally invasive endoscopic surgery was a rarity in China, and Zhou Pinghong was little known. It was an unexpected encounter that changed his life.

At that time, Zhou had wished to master more knowledge in his spare time. Professor Yao Liqing suggested that he study endoscopy. Ten years ago, endoscopy was mainly used for the examination of digestive tract diseases. Few would know that it would become a scalpel operation.

In the next few years, Japan applied the technique, and quickly established a firm position in the field worldwide.

In 2006, Zhongshan Hospital sent Zhou Pinghong to Japan to study endoscopic resection of early gastrointestinal cancer.

In Japan, Zhou studied with great diligence. The Japanese doctors got off work at 6 or 7 p.m., but Zhou Pinghong could keep working by the operating table from morning to 10 p.m. He knew that no skill could be acquired without hard work.

Zhou Pinghong, who was born in a village in south China, is familiar with the toil during the busy farming season. What he remembers most clearly is his parents' happy smiles after their hard work. He knows that it was a sense of fulfillment.

When Zhou Pinghong studied in Japan, he asked the most questions among all the students. More questions meant more ideas for solving problems.

By the time Zhou finished his study abroad, he had developed his own insights and extraordinary judgment concerning endoscopic surgery.

In 2010, Zhou not only successfully carried out the first endoscopic resection of early gastrointestinal cancer in China, but also created the POEM approach featuring invisible cavity. From then on, the endoscope, which used to serve only for examination purposes, became an extremely versatile scalpel in his hands.

After successfully performing several hundred POEM surgeries, Zhou Pinghong became better known internationally. In 2012, he was invited to attend the 14th International Endoscopy Symposium in Dusseldorf, Germany, representing China.

Zhou Pinghong was invited only to watch and learn. At that time, some European countries, America and Japan dominated the endoscopic gastrointestinal surgery stage.

But this time, China too had every reason to stand on the stage.

Zhou Pinghong has participated in many medical conferences.

After much effort, the congress agreed to offer China a chance, on the condition that Zhou Pinghong perform an operation at the same time as a top Japanese expert.

It was a competition about not only skills, but also dignity.

Zhou Pinghong went into the operating room. At that moment, he knew that academic circles had little trust in Chinese doctors. He had once actually heard that China was definitely not good at training surgeons.

In fact, doctors from Japan, Germany, the US, Russia and India had all stood in the center of the surgical stage at the congress, but no Chinese doctor had ever stood there.

Zhou Pinghong wanted to change the world's view of China and Chinese doctors.

After only half an hour, Zhou finished the operation almost flawlessly. By that time, the Japanese doctor had not removed the nidus.

His almost flawless surgery amazed the world. A German doctor who watched the surgery in the operating room still well remembered the scene three or four years later. He said it was "a miracle, an incredible miracle. The Chinese doctor was amazing."

Zhou Pinghong's performance on the world's professional stage in 2012 demonstrated the efficiency, quality and skills of Chinese doctors in endoscopic surgery, and established China's leading position in minimally invasive digestive endoscopic resection.

Professor Yao Liqing remembered that he shed tears at that moment.

Those of great aspirations never stop in their pursuit. Gaining global recognition was only an initial success. Zhou Pinghong extended application of endoscopic surgery, which had been confined to the gastrointestinal, thoracic and abdominal cavities. This was an epochal breakthrough, for more patients could benefit from minimally invasive endoscopic surgery apart from those suffering from gastrointestinal diseases.

This was a milestone for endoscopists all over the world. In the future, endoscopic treatment will break through more restrictions and extend beyond the gastrointestinal tract.

Apart from improving his surgical skills, Zhou Pinghong dedicates himself to the development of domestic endoscope auxiliary tools, including a suture device.

In some gastrointestinal endoscopic operations, the wall of the gastrointestinal tract has to be penetrated, and the wound needs to be sutured after the operation. In the past, foreign suture devices which only emit one metal clip at a time were used. Therefore, one suture process required replacing multiple suture devices, which meant great pain for the patients, more work for the doctors and high surgical risks.

Zhou Pinghong wished to develop a suture device of multiple metal clips so that there would be no need to replace suture devices and the surgical risk would be greatly reduced.

And, most importantly, he believes that once such a suture device is available, the cost of the surgery will be remarkably reduced.

It is an ingenious idea, just like his Zhou's surgery. Zhou Pinghong believes that there can be no innovation, nor progress, without ingenious ideas.

The starting point for doctors is to provide the maximum guarantee for patients' lives through medical technology. The essence of doctors' benevolence is love. Zhou

Pinghong strives to solve various problems in endoscopic surgery to bring patients back to a healthy life. He wants to benefit more patients with his love.

Great artisans all love and desire to benefit others with their expertise. It is also their fundamental source of power to solve the most difficult problems and their greatest value. Zhou Pinghong, an excellent doctor who genuinely cares for his patients, is one of them.

A Doctor of Dedication and Benevolence

In the spring of 2016 it finally got warm in Shanghai after a particularly cold winter.

In that year, there was a heated argument on the differences between Chinese and German and Japanese craftsmen. By 2016, the *Great National Craftsmen* of CCTV had been broadcast continuously for several years, and May Day was no longer just a holiday for sightseeing and relaxation. A lot of people gradually got used to talking more about China's industry, and Chinese workers and craftsmen.

The Internet economy is booming in China. A lot of people, including me, hold that the world will develop in that direction. So would there be much chance for industry?

It was with such a question that I went to Shanghai by myself for the preliminary research of the program. In Shanghai, I had two choices for the protagonist of my program: Wang Wei, a sheet metal worker of COMAC, and Zhou Pinghong, an endoscopy expert of Zhongshan Hospital.

When I left COMAC and headed for Zhongshan Hospital by subway, I kept thinking about a sentence that I had read somewhere: A doctor is actually an artisan.

In the ancient times, doctors were defined as craftsmen in China.

When I came out of the subway and stood in front of Zhongshan Hospital, I felt a bit of the magic atmosphere of the magic city of Shanghai. The roads were not wide and there was a lot of traffic, yet I didn't feel squeezed. Shanghai is always exquisite in many respects.

But I didn't see Professor Zhou Pinghong. He was giving lectures in other cities.

Yet my trip was rewarding. Professor Yao Liqing, the founder of the Endoscopy Center of Zhongshan Hospital, received me in his small office.

It was in this office that I learned about the history of endoscopy in China.

Too difficult!

These two words were sufficient to summarize the history.

Before I had time to look at the photos on the wall of the Endoscopy Center, Professor Yao Liqing began to tell me about Professor Zhou Pinghong. Through his narration, vivid images of Zhou Pinghong were conjured up in my mind.

In a rice paddy in south China, a boy was running towards his parents. He stopped and smiled at them; his feet covered with fresh mud. It was noon time, and they returned home together. Outside the farmhouse, the boy picked up a thick book to read, and a small dog came to play with its little master . . .

The boy grew up, and embarked on his long journey of study, in Shanghai, Hong Kong, Japan . . .

Professor Zhou will certainly not blame me for having such an impression of him at the beginning. In our contacts later, he often mentioned his hometown, his eyes full of yearning.

Zhou Pinghong is both a professor of profound knowledge and a highly skilled surgeon. As for me, I was most interested in his scalpel.

Professor Yao got up, and took me to the medical equipment room. Unwrapping a brand-new package, he showed me the endoscope inside. It seemed to be an ordinary endoscope, except that there was a wire less than 1 cm long and as fine as a hair at the tip of it.

It was the scalpel.

A few months later, I went to Shanghai again to see Professor Zhou Pinghong.

Because Professor Zhou had too many patients, our meeting was so hurried that we barely had time to greet each other. In his routine rounds, Professor Zhou strode ahead so rapidly that I could hardly keep up with him.

Professor Zhou Pinghong, middle-aged, always had a big smile on his face, which was quite reassuring to the patients lying in bed. He seemed to remember every patient, even if he had not performed an operation on that patient. He told a couple from rural Jiangsu Province that their daughter could take liquid food, and told an old man that he could go home the following day . . .

After the round, the atmosphere in the ward became lively. It must have come from the cordial trust of the patients in Professor Zhou, which shortened the distance between them.

Professor Zhou spent the rest of the day performing operations.

There are no visible wounds in endoscopic surgery; all the procedures are performed inside the patient through the probe of the endoscope.

Among all the figures in *Great National Craftsmen* I have seen, Zhou Pinghong is the only one who works without relying the touch sense of his hands. He controls the other end of the long endoscope, leading the hair-like scalpel to cut the nidus with great precision.

It is needless to describe again Professor Zhou's superb skills here. I just want to say one thing: I stood there watching him perform four surgeries in a day in astonishment and admiration.

What made me more astonished was that as endoscopic surgery is done inside the patient's body, the operation environment is relatively open, so I practically watched the operation process amongst a crowded audience.

Every day, the Endoscopic Center of Zhongshan Hospital is crowded, not only with patients, but also with endoscopy doctors from all over China and other countries.

As soon as Professor Zhou Pinghong put on his surgical gown, many eager learners, all famous doctors, surrounded the operating table, making the narrow operating room of less than 10 m² even more crowded. In the first surgery, my view was often blocked by many heads in front of me; they were more eager to see the details on the screen than I was.

There are people who are always racing against time in life. I've been fortunate enough to spend some time with some of them.

After five consecutive days of performing surgeries, we followed Professor Zhou to Nanjing on a Saturday morning. In Nanjing, he was to attend an academic discussion at 8 o'clock and give surgery demonstration at 10 o'clock.

I can't quite remember the names of the two Japanese endoscope experts. They used to be among the best endoscopic doctors in the world, but things are different today.

Professor Zhou was humorous. He told the Japanese doctors that they could watch, but they could not film. On hearing this, the Japanese doctors were surprised, but Professor Zhou laughed and said that he was only joking, that it was OK to film and he was willing to answer their questions to the best of his knowledge. The two Japanese doctors were relieved and excited; there was no reserved air of accomplished doctors about them.

Professor Zhou was always busy. I could hardly see him take any rest.

We followed Professor Zhou and his team to Greece, where the Athens International Symposium – Gastrointestinal Caner was to be held. The conference, which was known as the Olympics of endoscopic surgery, had been dominated by Western and Japanese doctors at one time.

But the Chinese team came, with Professor Zhou at its head.

The wife of the president of the congress spent nearly a year painting a portrait of Zhou Pinghong.

Before the conference started, I discovered that many people hastened to greet Professor Zhou, and almost everyone was proud to know him. I was deeply touched.

Professor Yao Liqing once told me about a conference he and others had attended a few years previously. At the three-day conference, not a single minute was given to Chinese doctors. Pride and prejudice did exist at that time. But the Chinese doctors who dared to attend the event were well prepared: to communicate more with and learn more from peers and benefit patients with what they learnt.

The Chinese doctors did not take the conference as a decisive battle, nor did they intend to conquer with one stroke. They were just eager for an opportunity to watch expert surgical demonstrations, and they were given the opportunity. Their demonstration of surgery became a spectacular event of the year. Chinese doctor Zhou Pinghong completed an extremely complex operation with the utmost accuracy in the shortest time!

At that moment, the world re-learnt about China.

When the conference was held in Greece, Professor Zhou Pinghong performed the first operation in a local hospital. A video of the operation was transmitted to the conference room simultaneously.

I still remember the complete silence in the conference room containing several hundred participants. I believe I will always remember, many years later, how Professor Zhou Pinghong heading towards the operation table with firm steps as the door to the operating room opened.

Perhaps Professor Zhou Pinghong was used to the applause, as was his student Dr. Cai, a doctor who was quick and neat in her work. Applause and honors were commonplace to them.

But how could I be unaffected? I was so excited that I wanted to toast with someone for celebration.

Professor Zhou is busy, and so are his colleagues. They are carrying out programs to aid Tibet, Xinjiang and other parts of China. They have no reservations. They hope to improve the standard of endoscopic surgery in China, so that all patients can share the benefits of the world's most advanced surgery.

They have been working hard with deep love for the country and its people.

It was lucky that Professor Zhou's episode of *Great National Craftsmen* was broadcast on National Day in 2016.

This article was written two years later, in 2018. After just two years, many people, including me, are still having heated discussions about the revival of industry. I have become friends with many artisans.

I will always remember the moments spent with Professor Zhou Pinghong and his family (his wife Zhang Ying and his son Zhou Yuhao), Professor Yao Liqing, Head Nurse Wang Ping, and Professor Zhou's students Dr. Li Quanlin, Dr. Cai Mingyan. I will always cherish those exciting moments that made me shed tears. Thank you!

I also want to thank my colleagues Jiang Li, Liang Zhiqi, Zhang Wenkai and Wang Wenbo. You are like my brothers and sisters, and I appreciate the days when we struggled together.

By ZHANG YONGFENG, documentary director

SHAN JIAJIU

Working with Eternal Conscience

Profile

Shan Jiajiu is an inheritor of mounting and restoration techniques applied to ancient paintings and calligraphic works. On December 28, 2017, Shan Jiajiu was put on the recommendation list of major inheritors of the fifth batch of representative national intangible cultural heritage projects.

Shan Jiajiu has been working in the science and technology department of the Palace Museum since she was 21. So far, she has restored more than 200 precious paintings and calligraphic works. Basically, ancient painting and calligraphy restoration includes four steps, i.e., washing, uncovering, repairing and completing. The restoration of each item is a long cycle, the longest lasting several years. In this process, a restorer can never slack off, for the relics will be easily damaged even by the slightest carelessness. Being bold but careful is the most important quality for a restorer.

Shan Jiajiu's father, Shan Shiyuan, used to be deputy president of the Palace Museum. In the "warlord era" (in the 1910s and 1920s), Shan Shiyuan guarded the first batch of cultural relics evacuated from the Palace Museum. He devoted his entire life to cultural relics. He died at the age of 91 and was the only one who had worked in the Palace Museum his entire life. He made a rule for his family that if any family member's work was related to cultural relics, he or she should not collect or deal in them. Shan Jiajiu followed her father's rule. She lives a simple life, and is totally dedicated to painting and calligraphy restoration.

Shan Jiajiu, painting and calligraphy restorer of the Palace Museum

Preview

Painting and calligraphy restoration is a time-honored craft. Today, despite the advances of science and technology, the ancient paintings and items of calligraphy that have been passed down can only be preserved via this traditional craft. A large number of original paintings and calligraphic works displayed in the Palace Museum have been seriously damaged over the centuries, so restoration is conducted to extend their life. One bout of restoration can extend the preservation period for over 100 years. There is a group of top restorers in China's Palace Museum who specialize in the restoration of the first-class precious cultural relics. The 59-year-old Shan Jiajiu has been engaged in this work for 38 years.

The Science and Technology Department of the Palace Museum is located in a secluded alley.

Shan Jiajiu goes to work on foot every day. It takes her 50 minutes from her home in the Drum Tower area to the Palace Museum. She has kept up this habit for 38 years.

After she crosses Shichahai Lake area, the Palace Museum leaps into her gaze immediately. Today, the magnificent palace complex built in the Ming and Qing dynasties (1368–1911) greets tens of thousands of tourists from all corners of the world. Shan Jiajiu then enters a secluded and obscure alley.

The Science and Technology Department, where Shan Jiajiu works, is located on the west side of the Palace Museum. The building, covered in flourishing crabapple flowers, is said to have been an isolated building in the Qing Dynasty (1644–1911).

The door of the restoration room is double, one layer being added later for wind proofing. Shan Jiajiu told us: "The paper wall opposite the door is used for hanging the paintings and calligraphic works, and keeping them even. In spring, the wind in Beijing is very strong, plus the big pulling force generated by the paper itself, the paper will be easily torn apart. So we must take great precautions against the wind." There are only two keys to the door. Whoever comes first fetches the key to open the door, and whoever leaves last closes all the windows, locks the door and returns

117

the key to its proper place. "In any case, the key should never leave the Science and Technology Department."

There are tremendously valuable cultural relics in the Palace Museum, and Shan Jiajiu is familiar with the preferences of each emperor in painting and calligraphy. For example, Emperor Qianlong (1736–1796), who was also a master of calligraphy and painting himself, would issue edicts to instruct his painter how many cranes or pine trees should be put in a painting.

The ancient paintings and calligraphic works in the Palace Museum needing restoration fall into two categories. One is mainly cultural relics collected by royal courts and handed down to later generations, including calligraphy, painting, rubbings of stone inscriptions, and portraits of emperors and empresses. Examples in this category include *Bo Yuan Tie* (a calligraphic work by the Eastern Jin Dynasty calligrapher Wang Xun), *Wu Niu Tu* (a painting of five oxen by Han Huang of the Tang Dynasty) and *Riverside Scene at the Qingming Festival* by Zhang Zeduan of the Northern Song Dynasty. The other category covers royal relics, including plaques, calligraphic works and paintings of emperors, royal officers and artisans. These antiques would not survive long without the restoration techniques.

This day, Shan Jiajiu and her apprentices were to restore a paste-down landscape painting which was originally displayed in Jingqi Pavilion.

Paste-down paintings were a common decoration in royal courts, executed by either emperors or royal artisans. In recent years, the Palace Museum has been restoring the original appearance of several halls. In order that visitors can appreciate the genuine royal charm of ancient times, damaged paste-down paintings will be restored and put back in their original places. Shan Jiajiu is a member of the painting restoration team undertaking the task.

To our surprise, Shan Jiajiu's first step of restoration was to wash the painting with hot water. This was because the colors of the ancient paintings were fixed and would not be washed off. Hot water was used because the dust the painting had accumulated over so many years that it could not be washed off with cold water.

The next step was to remove the several layers of mounting paper to reveal the thin piece of rice paper on which the painting had been done, so as to repair the damaged areas.

Shan Jiajiu washes the painting with hot water.

She carefully removed the first layer of backing paper, a Korean paper called Hanji made in Emperor Qianlong's reign. The backing paper, with a history of several hundred years, could be used as repair material in the future.

Shan Jiajiu said that Hanji paper is strong and durable. She had tried to reproduce this type of paper in joint efforts with a paper mill, but they couldn't produce any paper of the same quality.

After removing two layers of backing paper, it was time to remove the last layer immediately beneath the painting. This was the most difficult and the most critical step in the entire repair process, and any mistake would damage the art piece. Although Shan Jiajiu has rich experience of 38 years in painting and calligraphy restoration, she is always extremely careful at this step, as if she is treading on thin ice.

As the paste-down landscape painting that Shan Jiajiu was repairing was very old, the supporting paper was rotten and couldn't be separated from the painting. In order to ensure the integrity of the painting, Shan Jiajiu decided to adopt the rubbing method, i.e., rubbing the supporting paper off bit by bit. It was an arduous and time-consuming task.

Shan Jiajiu removes the supporting paper of a painting.

The supporting paper was closely attached to the painting, and the thickness of the two layers of paper was only 0.22 mm. Therefore, it was quite a task to separate them. It took years for a restorer to have fingers sensitive and nimble enough to do this job.

The ancients described the restoration of painting and calligraphy works as "seeking treatment for a serious illness." As the saying goes, "An excellent doctor cures, but a mediocre doctor kills." The ancients even proposed to "leave the damaged art pieces as they are unless there is a skilled restorer."

Shan Jiajiu's apprentice Yu Li, a graduate student of the Central Academy of Fine Arts, has been assisting Shan for two and a half years. Until now, Shan has not allowed him to carry out any restoration independently, but just asks him to do chores like mixing paste.

Yu Li said that although mixing paste sounds simple, there are actually many important details. Different processes require pastes of different viscosity.

It was a rule passed down by Shan Jiajiu's supervisor that apprentices should not actually repair cultural relics for three years. When Shan Jiajiu was young, she spent a long time practicing the basic skills of making brushes, scraping paper, etc.

Shan Jiajiu restores an ancient painting.

One thing Yu Li admires is the way Shan Jiajiu can often tell if he makes a mistake by merely listening. For example, if he is using the brush with appropriate strength, if his operation can meet the requirements, etc.

Yu Li said that his biggest gain was not skills, but his appreciation for the reverence of the senior experts towards cultural relics. Despite her rich experience of 38 years in restoration of cultural relics, Shan Jiajiu never dares slacken off but always works with undivided attention. The senior restorers often say that cultural relics must only be repaired, and never damaged.

It took Shan Jiajiu three days to remove the supporting layer of the drawing. After pasting and mounting, the painting looked just as splendid as it looked three hundred years ago.

Then she proceeded to patch up hundreds of holes, large and small, that had been caused by worms and natural wear on the painting. It would take three months to restore it completely.

I once asked Shan Jiajiu: "What is the most famous and valuable art piece that you have repaired?" Shan replied slowly: "To a restorer, all art works are the same. We are called 'art work doctors' because our relationship to art works resembles that between doctors and patients. For a patient, what medicine or treatment shall be taken depends on the disease and the patient's physical condition. When art works are 'ill,' how to rescue and restore them depends on the state of damage rather than the status or value of the art works. In this sense, from the national treasure of *Wu Niu Tu* to an ordinary paste-down painting like this one I'm currently working on, the difference lies only in the difficulty of restoration."

Shan Jiajiu told us that the restoration of cultural relics is a work more based on conscience. Outsiders can hardly tell with the naked eye how an art work has been restored, but an insider knows exactly the inner secrets.

The most difficult work to repair was *Shuang He Qun Qin Tu*, a painting of two cranes and a flock of birds of the Ming Dynasty. The painting, which was done on a silk scroll two meters long and nearly one meter wide, was badly damaged with numerous small holes.

Restoring a silk scroll is much more complicated than restoring a painting on rice paper. The easiest method, called whole-piece repair, involves applying a whole piece of silk to the back of the painting to cover all the holes, with further slight adjustments. As Shan Jiajiu is highly skilled, she can deliver a perfectly restored work with this method in one week.

However, if a painting is repaired by the above method, the whole piece of silk will rot as time passes, for example, after 100 years. By that time, the silk will stick to the painting itself, which will no longer be restorable.

After weighing the advantages and disadvantages, Shan Jiajiu decided to repair the small holes one by one.

It took her more than four months to fix the holes one by one. Now the back of the painting is densely covered with over 1,000 small patches.

Shan Jiajiu said that the restoration of cultural relics is a work of conscience. She will never allow the art works to be damaged under her hands. She is determined to restore the art pieces and pass them down to future generations. This is the duty of cultural relics technicians.

Shan Jiajiu's husband is an expert in the restoration of ancient buildings in the Palace Museum. Her father, Shan Shiyuan, was deputy head of the Palace Museum. At the age of 17, Shan Shiyuan, as a member of the "Final Work Committee of the Imperial Court of the Qing Dynasty," entered the Palace Museum to make an inventory of its cultural relics. In the chaotic years when warlords coveted the treasures of the Palace Museum, Shan Shiyuan guarded them with his life. Ever since then, he never left the Palace Museum until his death at the age of 91. He was the only one serving the Palace Museum his entire life. "Whereas my father was fond of appreciating the magnificent architecture of the Palace Museum, I have been touching its art pieces all day long for years, including hand scrolls, vertical scrolls, calligraphic works, paste-down paintings, fan coverings, etc." said Shan Jiajiu.

The lives of Shan Shiyuan and Shan Jiajiu are linked to numerous precious cultural relics, but there isn't a single antique in Shan Jiajiu's home. Shan Jiajiu said her father set the family rule that no one in the family should collect or deal in cultural relics if his or her work was connected with cultural relics.

Throughout her life, Shan Jiajiu has been living in the same place and doing the same job. Once a German museum offered her a well-paid job, but she declined it. She has been observing strictly her father's teachings throughout her life. No matter how hot the cultural relic market is, she has never engaged in it. For more than 30 years, she has been restoring national treasures one after another in the small workshop of the Palace Museum.

The Palace Museum has about 150,000 items of paintings and calligraphic works, which account for about a quarter of the total collection of Chinese paintings and calligraphy in public museums throughout the world. Among the collections there are a number of top-level cultural relics or national treasures, such as Zhang Zeduan's painting *Riverside Scene at the Qingming Festival* and Gu Hongzhong's *Banquet of Han Xizai*, which have been restored by Shan Jiajiu and her colleagues. A large number of paintings and calligraphic works which have been exhibited in their original places in the Palace of Museum are in urgent need of restoration. There are so many works to be restored that generations of artisans will be needed.

Shan Jiajiu will retire in one year. Having spent so many years in painting and calligraphy restoration, she has blurred vision, and her legs and knees ache due to long hours of standing. Many "painting doctors" in the Palace Museum are hired back after retirement because there are so many art pieces waiting for them. Shan Jiajiu's biggest wish is to pass on her experience in traditional painting and calligraphy restoration to the next generation, and, together with her colleagues, restore the cultural relics of the Palace Museum in a perfect way and hand them over to future generations.

The videographer shoots the working scene of Shan Jiajiu.

Approaching the "Conscience" of a Great Master

"Shan Jiajiu: Working with Eternal Conscience" was one episode of the third season of *Great National Craftsmen* broadcast in 2016. The previous two seasons had gained much public praise and exerted a profound social influence. Therefore, it was challenging to be innovative and make more breakthroughs in the third season. We were all under great psychological pressure.

I would like to summarize the features of the third season with the word "dedication" that ran through the interviews, shooting and editing of each episode. All the subjects of *Great National Craftsmen* have superb skills, but what makes them all the more distinguished is their tireless pursuit of excellence. It is their drive for excellence that enables them to attain remarkable achievements. Therefore, this season we strove to explore the inner world of the characters to render the images of the great national craftsmen in a more vivid and three-dimensional way.

Understanding the great master

While the Palace Museum is always crowded with tourists, there are some seclusive corners not open to the public, as quiet as if secrets have been kept there undisturbed for hundreds of years. Shan Jiajiu has been working in one of the small yards of the Palace Museum surrounded by crabapple trees for 38 years.

I observed Shan Jiajiu's daily life. In the morning, she goes to work on foot at a leisurely pace. A century-old painting is spread out on her work table, which she repairs bit by bit and day after day. The entire restoration may take several weeks, and she carries out her work systematically and methodically. After one painting is finished, she starts to repair the next one.

In making a program, we do our best to avoid monotony, but Shan Jiajiu's life has been almost the same for 38 years. I asked her if she had to work overtime, or if there was any performance evaluation each month. She shook her head, and told me with a smile: "It is up to me how much time will be spent on a painting. No one pressures me. If I work in a hurry, I might make a mistake and damage a painting. I can't rush, nor work overtime, because the Palace Museum closes at six o'clock every day, and no one is allowed to stay after work." On hearing this, I couldn't help drawing the conclusion that it was a comfortable job. But how could I represent it in my episode?

Many things in the world resemble the sea, which has a calm appearance but a wonderful world hidden beneath the surface. After making the previous two seasons of *Great National Craftsmen*, I knew I had to find the key to discovering a new world.

The key to Shan Jiajiu's inner world was a sentence she said inadvertently to her apprentice. It was a breezy afternoon in spring, and a few elderly restorers were taking a nap in the studio. Shan was repeating the operations that she was so familiar with. She seemed as calm and casual as usual, but she said to her apprentice that she felt like "treading on thin ice." The phrase sounded out of harmony with the environment and formed a sharp contrast with the tranquil atmosphere in the studio. As if enlightened by the words, I immediately grasped the essence of her work.

This small, seclusive yard, which is separated from the noisy world outside, seems extremely tranquil and comfortable, but the people working here are definitely not as detached nor as unrestrained as they seem to be. On the contrary, they are highly responsible and attentive in their work. The antiques they restore have survived so many years, and are all unique and extremely valuable. In Shan Jiajiu's words, the most important thing for them is to hold a reverent feeling towards the cultural relics.

We carried out the interview following this direction, which lead to Shan Jiajiu's story of repairing the painting of two cranes and a flock of birds. The restoration took two months. Shan could have taken the method of "whole-piece repair" by using a piece of silk to cover the more than 4,000 small holes on the back of the painting at one time. It would only take her one week, and the result would seem to be perfect. However, she chose to repair the small holes one by one, which took her four months. Shan explained that the quality of restoration of art pieces could not be immediately tested, but only revealed several hundred years later, when later generations would check the restoration. For the painting in question, it would not make any difference to the current generation if she simply adopted the "whole-piece repair" method. However, when this painting needs to be repaired again several hundred years later, it may be difficult or even impossible to restore it, and the painting might disappear from the world. Shan Jiajiu said she did not want to be such an irresponsible restorer and be blamed by future generations.

It is truly a work of conscience. No one makes demands, nor supervises her work. It's all based on the repairer's "reverence" for cultural relics and sense of responsibility.

At that moment, I understood Shan Jiajiu's "conscience." She has lived in the same place and been doing the same work for 38 years. In the world outside, full of temptations, many people tend to seek shortcuts, but Shan Jiajiu has chosen to restore antiques for future generations. She said that in some cases, the single step of "putting a painting up on wall and flattening it" would take one year. In this period, she would have to take good care of the painting just like a mother taking care of her baby, for the painting should be neither too dry nor too wet. Perhaps because of this simple requirement, Shan Jiajiu doesn't feel that "earning a lot of money" is attractive to her. She is totally content with her quiet life surrounded by cultural relics.

Breaking new ground for innovation

The episode of Shan Jiajiu is in the third season of *Great National Craftsmen*. The previous two seasons, which were made with elaborate and meticulous efforts, were of high quality. In the third season, we didn't want to repeat the previous seasons, but strove for new features.

Before the shooting, Jiang Qiudi, director and chief planner of this season, set new requirements in terms of the style of the program. In terms of narrative mode, we didn't adopt the commentary common in news programs, but picture language, documentary footage and the synchronous self-statement of the protagonist instead. By doing so, we wanted to approach the inner world of the protagonist. Through the narration of the subject, the narrative perspective was actually changed from the objective perspective of the journalist to the subjective perspective of the protagonist of the program, who "communicates" directly with the audience and better approaches the audience via this heart-to-heart talk.

However, this approach meant tremendous challenges to the producers. The episode was only about eight minutes, but it carried so much information, i.e., the particular background and specialized skills of the protagonist. Could the audience understand the program without commentaries? The path to innovation is full of unknown challenges.

One may well reach a new land, but one has to trudge through every underground river before reaching the destination.

Without commentaries, many scenes could only be expressed through camera language and documentary footage. The audience, instead of being told directly by journalists, were supposed to feel everything in the program by themselves. This greatly increased the difficulty of our shooting.

In fact, we adopted the shooting techniques of documentaries to strengthen the sense of reality and capture the wonderful moments in daily life. For most of that week, our cameraman Duan Dewen and I stayed quietly in Shan's studio, watching her closely to choose the timing of our shooting. We abandoned our former approach of directing our interviewees with a microphone. Instead, we were like leopards hunting, not disturbing the prey but watching and waiting for the opportunity to pounce.

For example, Shan kept telling her apprentices the key points in the entire restoration process. This was very good documentary material for us. She mentioned that the feeling of using a tool was "to be able to communicate with the tool." She told them that the first layer of backing paper, the precious Hanji paper made in Emperor Qianlong's reign, should not be abandoned as it could be used in future restoration of other works. Moreover, she admitted frankly that she felt as if she was "treading on thin ice" in the key steps of restoration. In these common and natural scenes, everyone was relaxed and real. Through these details, we could feel Shan's dedication and pursuit of perfection in her work.

In addition, as we could not rely on commentary in the narration, we had to force ourselves to find visualization means, which were actually closer to the nature of audio-visual communication.

For example, Shan's father set the family rule that if a family member's work is related to cultural relics, he or she cannot keep or deal in them after work. Therefore, there wasn't a single antique in Shan Jiajiu's home. As we couldn't present this detail in narration, I decided to present it in another way, i.e., by asking questions. In the program, I pointed at a painting in

Shan's house, and asked her: "Is this an antique?" On hearing this, Shan and her husband couldn't help laughing. The couple told me it was a replica. Shan's reply and expression at that moment reflected her simple life more vividly and persuasively than any narrative.

Making program with great dedication

Since there was no commentary in the program, the narration was completely carried out via images, which required a clear and neat rhythm for the visual display. This posed a great demand for the film editing. The process of film editing was definitely not matching scripts with images; it involved re-creation of the materials. It was not enough to display the protagonist's external behavior, but all the more important to reveal her inner world, for example, through her gestures and expressions.

For the first time in my career, I felt that editing a program was like practicing Chinese shadow-boxing. I had to follow closely the fresh feelings that touched me, and follow the rhythm of my thoughts. The duration of footage should be neither too long nor too short; it should be strictly controlled to realize the best effect. My editor Niu Xiaochen and I worked almost nonstop for more than 90 hours on the editing. We thought over every editing point and did our best not to take liberties. I was exhausted both mentally and physically.

Our hard work finally paid off. After the editing and re-creation, trivial bits irrelevant to the theme were removed to make room for more important parts, the program was neater and would better strike a chord with the audience. The narrative rhythm of the program was slowed down, giving the audience room to savor the inner world of the protagonist. Thereby, the program avoided the rigid pattern of "introducing deeds" and became really touching.

Great National Craftsmen has been running for six seasons. In this time, our pursuit of quality has never changed. Shan Jiajiu's concentration on her work inspires me to be more attentive to my own work. I will treat every interview, every scene and every footage in my program with utmost responsibility and more drive for perfection to present excellent programs.

By ZHANG QIANQIAN, reporter of
China Media Group & CCTV News Center

NING YUNZHAN

High-Speed Rail Technician

Profile

Ning Yunzhan, born in March 1972, a member of the Communist Party of China, is a senior vehicle fitter of CRRC Qingdao Sifang Co., Ltd. Over the past 26 years, Ning has been engaging in the grinding and assembly of bogies of high-speed bullet trains. With superb skills and high sense of responsibility, he has broken through bottlenecks in the manufacturing of domestic high-speed bullet train bogies, and has made outstanding contributions to the manufacturing of high-quality bullet trains. As of May 2018, he had kept an 11-year zero-defect record. The bogies he and his team produced are used on more than 1,300 trains running a total length of over 2.3 billion kilometers, equal to over 50,000 circles of the earth. The research topics he presided over and the work uniform he designed can save more than 3 million yuan for the company every year. Ning Yunzhan has won a series of honorary titles, such as "National Ethical Model," "Excellent National Labor Model," "National Best Worker," "National Professional Ethics Pacesetter," "Model SOE Worker" and "Star of Shandong: Annual Top Ten Worker."

Ning Yunzhan, senior technician of CRRC Qingdao Sifang Co., Ltd.

Preview

In July 2015 Chinese President Xi Jinping visited CRRC Corporation, and commented: "China's high-speed rail and bullet trains are a shining business card of 'Made in China.'" In particular, the 380A train has received an independent intellectual property rights certificate from the US Patent and Trademark Office. In 2010, the 380A set a speed record of 486.1 km/h on the Beijing-Shanghai high-speed railway. Chinese leaders often cite the 380A to promote China's high-speed trains on their trips abroad.

These remarkable achievements represent the glorious progress of Chinese railway workers in promoting science and technology, as well as the superb skills of a technician who has devoted half his life to this endeavor. His name is Ning Yunzhan, a fitter and senior technician of CRRC Qingdao Sifang Co., Ltd. Probably the young man from Qingdao, who graduated from a railway vocational school, had never dreamed that he would become China's top high-speed train technician by producing the bogies for the 380A. By May 2018, he had kept an 11-year zero-defect record. The bogies he and his team produced are installed on more than 1,300 trains running a total length of over 2.3 billion kilometers.

I t seems that we haven't heard the word "craftsman" for very long. Especially in an age when new things keep popping up endlessly, the word "craftsmanship" seems to indicate being conventional and boring. Are there any craftsmen around us? Does anyone want to be a craftsman? The answer seems vague.

On December 3, 2010 China's CRH380A train, with independent intellectual property rights, set a world record of 486.1 km/h on a test run on the Beijing-Shanghai high-speed railway. When Premier Li Keqiang visits foreign countries, he will take along a model of the CRH380A to introduce China's high-speed railway technology to the rest of the world. The CRH380A has become a shining international business card of China's high-speed rail.

For the filming of this episode of *Great National Craftsmen*, the railway authorities recommended several interviewees, all excellent technicians dedicated to the development of China's high-speed trains. Today, some of them have assumed leadership positions, and some have become designers and no longer work on the production line. While reading through the recommendation materials, we were impressed with the following description of Ning Yunzhan of CRRC Qingdao Sifang Co., Ltd.: "He is the top grinding technician of CRH380A bogies. He has been working in the basic-level workshop up to now."

When our team arrived at CRRC Qingdao Sifang Co., Ltd., Ning Yunzhan was not quite at ease in our meeting, for he was not used to the office environment. Since he graduated from the technical secondary school, he has been working in his workshop for 24 years, and has seldom entered an office.

Our first impression of Ning Yunzhan was his accent, which was not quite standard. But when it came to technology, we immediately discovered that he wasn't an ordinary guy, but an absolute "technology genius." His biggest feature was his simplicity and sincerity. Isn't this the unique temperament of craftsmen?

Later, we interviewed Ning in his workshop, where he became immediately at ease and confident. He looked at the steel and iron parts of his lathe in the way an affectionate father would look at his children. He touched the parts gently, as if they all had life.

I threw out my most pressing technical questions: "What is a high-speed train bogie? Why does it need manual grinding?" The question suited Ning Yunzhan exceedingly well, and he explained it to me at length. He compared a high-speed train to a long-distance runner, stating that the wheels were like a runner's feet, the

Ning Yunzhan has been working on the production line for many years.

bogies were like his legs, and the positioning arms served as the ankles. A person with a sprained ankle couldn't walk. The bogie was a critical part serving a similar function.

"The wheel sets are fixed to the positioning arm through insertion at the nodes, and then the positioning arm is fixed with bolts. The roles of the node, which is fixed on the positioning arm, is to ensure that the wheel sets are securely attached."

One bogie of a high-speed train weighs 1.1 tons. The positioning arms rest on the nodes of the four wheels, with each contact surface being less than 10 cm². When the train runs at 300 km/h, the impact force on the contact surface is as much as 20 to 30 tons. If the gap is too big, the wheels may come loose. If tightly welded, the bogie could not be accessed again, which would affect the maintenance of the train. This is exactly the challenging aspect of the technology. Ning Yunzhan pointed to the bogie beside him, and said that while other parts could all be processed mechanically, the bogie required manual grinding, which was not only the case in China, but also that of the entire world in high-speed train production. According to international standards, the space reserved for manual grinding is only about 0.05 mm, or the radius of a hair. Without enough grinding, the bogie can't fit. But if ground too much, the motherboard worth more than 100,000 *yuan*

Ning Yunzhan grinds a bogie.

would be damaged and become useless. Ning Yunzhan's colleagues said that he was perfect in doing this job: He could "weave," as in embroidery, the faint vertical lines on the surface of the cut into a fine net with great frictional force. In the past ten years, Ning Yunzhan has been exercising his skill in this extremely narrow space, from where his technical authority and self-confidence spring.

Compared with being accepted by leaders, it seems more difficult to get praise from peers. Ning Yunzhan's colleagues agreed, "There are about 15 technicians in China who can handle a grinding space of 0.1 mm. But when it comes to 0.05 mm, currently Ning Yunzhan is the only one who can do it."

Ning Yunzhan works at Qingdao. More than 100 years ago, the Germans built a factory producing locomotives in Qingdao. Today, the Chinese have produced on their own the best trains in the world, and cultivated the best high-speed train technicians.

Ning Yunzhan, who is technically superb, has a minor weakness in communication. In the words of his apprentice, the master "is exceptional at work," but he is also "exceptionally taciturn off work!" Because of his introverted nature, his first wife left him. In the interview, he said, "I have spoken as much in these past few days as I did in the previous ten years, so how can there be more interviews? It is really tough working for a TV station!"

So I had no alternative but to communicate with Ning Yunzhan via text messages. Ning wrote that he did not like to talk; he just wanted to focus on his work. I wrote: "Take the interview, for instance. In my interview, your work is to explain what you do clearly. There are technical elements in my job, too – only you use grinding tools, while I use a microphone." Then Ning Yunzhan seemed to better understand me, and was more cooperative. He got used to talking to me wearing a microphone.

In fact, it was not my text message, but his six-year-old daughter that made him talk more in the interview. The little girl was lovely and talkative. Ning had wished to be as talkative as his daughter, but it was difficult to change his long-term habit. I asked him if he would like to have his daughter do the same job. He replied that he didn't have that plan as girls were not suitable for the job. Then he turned the topic to his father. His father, who was a blacksmith in their village, originally did not want him to be a craftsman, but a businessman and earn more money. Ning finally followed his heart, and chose a railway technical school.

The year 2010 was a crucial year for the 380A train, which had to undergo a high-speed test. In this year, Ning's father, who had suffered from leukemia for seven years, was hospitalized for the third time. Although Ning realized that there was not much time left for his father, he could not be at the old man's bedside every day. One day, when he was leaving work, he heard the bad news that his father had passed away.

"My family called, and told me that my father was gone. I hurried back home. I was very sad, because my father had had a great influence on me."

On December 3, 2010, *CCTV News* broadcast the news that China's self-designed 380A high-speed train had travelled at a speed of 486.1 km/h, on the Beijing-Shanghai railway. It was a pity that Ning Yunzhan could not share the good news with his father. With our parents alive, we have our origin of life. When our parents are gone, there is only destination. How sad it is to part forever from one's parents!

After that, Ning Yunzhan became even more devoted to his work. He said that as he had become a craftsman, he was willing to remain one throughout his life.

Ning Yunzhan examines a bogie that has just been assembled.

Ning also said that it was a blessing from fate that he had caught up with the era of the high-speed train.

In 2006, Ning Yunzhan was picked, from among numerous peers, as the first Chinese to learn the 380A train bogie grinding technology in Japan. The Japanese experts were impressed by Ning's perfect skills and precision control. In fact, Ning was somehow reluctant to learn from the Japanese at the beginning. Born in Qingdao, he had heard the older generation telling stories of Japan's invasion of this coastal city and other parts of China. He had secretly made up his mind that he would not let the Japanese look down on him. Sure enough, after months of hard study, and practice, he won praise from the Japanese. When talking about this, there was a big smile on Ning Yunzhan's face for the first time in the interview.

"The Japanese supervisor said via an interpreter that he would like to call me elder brother, for he considered me as a master and I was older than him," Ning said.

One must have faith. With his faith, Ning Yunzhan became the first and best technician capable of grinding bogies for the 380A train. Soon he became the leader of his team. However, not long after, he went to his supervisor, and said that he did not want to be a team leader, for he just wanted to work.

Many excellent workers in the production of high-speed trains had been promoted. I asked Ning Yunzhan why he was still doing the basic-level work as before.

Ning replied, "After the 380A train reached the high speed, I was appointed the team leader. But I soon asked to be allowed to return to the production line, because this is what I am good at."

I asked, "Isn't it better for you to be a team leader and engage in technical work at the same time?"

He said, "In that way, I would be a perfect person, wouldn't I ?"

I asked, "Don't you think you are perfect?"

He replied, "I am not, and will never be."

This is a craftsman of our great nation, who is obsessive about and excels in his work. His achievements are impressive: He is the first Chinese technician capable of handling the grinding work of high-speed train bogies. Not only did the foreign experts who taught him gave him praise, but also his apprentices are deeply impressed with and admire him. Today, there are fewer than 15 technicians in the whole of China who can do his job, yet he is the only one performing the most demanding part of the work.

Knowing Myself Better Through Interviewing Ning Yunzhan

Interviewing Ning Yunzhan was like reviewing my own experience. This feeling stems from my own life. We shared many similarities in our early years. We were both born in the countryside. In the year he graduated from junior high school, Ning Yunzhan chose a railway technical school, which was also my first choice. Unfortunately, I wasn't admitted. In the next more than 20 years, we had different experiences. When Ning Yunzhan is processing train parts in his workshop, I am doing interviews and revising my manuscripts. I had never expected that we would meet in a high-speed train workshop several years later.

From Ning Yunzhan, I have learned more about inheritance. Ning has a pair of dexterous hands like those of a magician, which he inherited from his father, a village blacksmith. When he was a child, he helped his father make furniture for the villagers, and became interested in craftsmanship at a young age. After graduating from junior high school, Ning was admitted to a railway technical school. Since then, his fate has been closely connected to trains. He has devoted so much time to his work that it is hard to separate his work from his life.

Ning Yunzhan's home is nearly half an hour's drive from the factory. He and his wife work in the same factory, but they mainly communicate with each other in their commuting time every day. When Ning arrives home, he is busy again. His stuff occupies much of the space of their small courtyard of only some 30 m². In order to improve his skills, he bought grinding tools online at his own expense. However, his wife didn't approve at first.

"After working for an entire day, he practices grinding until eight or nine o'clock in the evening. Aren't you tired? I asked him. Because of this, my own hobbies and interests have all disappeared. I gradually accepted his state of mind, and understood him. Different people think differently. To understand him and support his work, I must support all his thoughts." His wife said.

When I heard this, I fully understood Ning Yunzhan's wife. Understanding and acceptance are all marriage about. But his 6-year-old daughter is still too young to fully understand everything about him.

I asked her: "When you grow up, do you want to do your dad's work?"

"No, I don't."

"Why not?"

"Because it's tiring."

"So what do you want to do when you grow up?"

"Be an engineer."

While I was talking with his daughter, Ning Yunzhan was standing and operating his machine not far from us. I saw tears in his eyes reflected in my camera lens. Ning Yunzhan said that he would support his daughter, just like his father respected his choice when he was admitted to the railway technical school. His father hoped that Ning Yunzhan would become a technical expert who was irreplaceable to his work unit. After more than 20 years of hard work, Ning Yunzhan has finally made it.

But because of work, he failed to be at his father's death bed, which became his eternal regret.

"A craftsman should be honest and work hard. It is the duty of a craftsman to do his job well," he emphasized.

I asked him: "Till when?"

"Till I can work no more."

After the interview, Ning Yunzhan said that this was the first time that he had shed tears in front of a stranger. When I was writing this report late at night after the interview, I burst into tears, too. What an unusual experience for an economic journalist who deals with policies and figures every day! I appreciate that the program *Great National Craftsmen* offered me the opportunity to approach China's high-speed train and to meet Ning Yunzhan, a great technician. Through filming the program, I gained a deeper understanding of myself and my career.

The core value of *Great National Craftsmen* lies in its ingenuity and persistence. With this persistence, Ning Yunzhan has become a superb technician. This persistence will also drive me and my colleagues to forge ahead. A doctoral tutor of Communication University of China watched the program broadcast on *CCTV News*, and sent me the following WeChat message: "I have watched your program twice. It is very good in its news elements and human feelings, especially the part about the little girl. It is a breakthrough for *CCTV News*. I have downloaded the episode, and will use it in my class."

Three years have passed. The research topics Ning Yunzhan presided over and the work uniform he designed can save more than 3 million yuan for the company every year. He has been awarded a series of honorary titles, such as "National Ethical Model," "Excellent National Labor Model," "National Best Worker," "Model Worker of Centrally-administered SOEs," and "Star of Shandong: Annual Top Ten Characters."

The brand of "Great National Craftsmen" is still active today. New blood has kept joining in, and the program has received multiple rewards and honors, such as China News Awards and National Labor Medal. As for me, joining this team and shooting the craftsmen is not only a practice of professional skills, but also elevation of my mindset. Through filming the craftsmen, I have identified myself with them. Although I didn't have the opportunity to participate in the filming of each season of the *Great National Craftsmen*, my working experience of three seasons is a valuable asset in my career.

In shooting *Great National Craftsmen*, I met myself, and talked to myself through conversations with the craftsmen. Who am I, where do I come from, and where am I heading? What is news? How to explore and how to innovate? As old questions remain to be answered, new challenges are coming. However, I firmly believe that with the experience of filming and retrospection, we can move forward more confidently as the new era of media integration arrives. We shall not indulge in empty thoughts or empty words, but combine knowledge with practice and strive for more actual results in this new era of the press.

By ZHENG LIANKAI, reporter of
China Media Group & CCTV News Center

ZHU WENLI

Legendary Ru Porcelain Master

Profile

Zhu Wenli, born in Ruzhou, Henan Province, in 1946, is a prominent inheritor of national intangible cultural heritage, great master of Chinese ceramic art, recipient of a special State Council allowance and member of the Chinese Society for Ancient Ceramics.

Zhu Wenli has been engaged in the development and production of Ru porcelain since 1976. In 1987, he developed Tianqing glaze (celeste-blue glaze) from a Ru kiln. In June 1988, Tianqing glaze passed appraisal by the Ministry of Light Industry, and was included in the Chinese Technical Achievements of 1990, reviving Ru official porcelain, which had disappeared for several hundred years. In 1993, he presided over the key project of "Imitation of Official Ru Porcelain of Baofeng Qingliangsi Temple" designated and appraised by the Henan Province Science and Technology Commission. In 1994, he was awarded the honorary title of "Star of Invention, Innovation and Technology" by the UN Technological Information Promotion System. In August 1998, he won the gold award at the First China International Folk Art Exposition. He was invited many times to the International Forum on Ancient Ceramics Science and Technology, has published more than 120 papers and is renowned as "the leading master of celadon ware."

Porcelain master Zhu Wenli

Preview

Born in the middle of the Tang Dynasty (618–907) in Ruzhou City, and reaching its heyday in the Northern Song Dynasty (960–1127), Ru porcelain ranked the first of the "Top Five Porcelains." Eight hundred years ago, the Ru official kiln was destroyed in the war between the Song and the Jin dynasties. The craftsmanship was lost and the formula of the porcelain became a mystery. Currently there are only 65 pieces of ancient Ru porcelain extant in the world. For more than 800 years, countless artisans devoted their lives to recreating Ru porcelain, but they all failed. It was the ultimate challenge for ceramics artisans. But Zhu Wenli's persistence finally paid off. In order to find out the long-lost secret, he strained every nerve developing nearly 400 formulas. He worked day and night at kilns with temperatures of more than 1,000 degrees to make the celeste-blue porcelain he had been dreaming about. There ensued more than 1,000 tests, more than 1,000 disappointments, and more than 1,000 smashing of unsatisfactory wares. But Zhu never gave up. Five years later, he finally created the miracle, bringing Ru porcelain back to the world. The news shook the industry, and he became "the first modern Ru porcelain master." Despite this success, Zhu did not relax his efforts to bridge the gap between himself and the ancient artisans. He has devoted his life to looking for the key ingredient for his formula after more than 40 years' research. Now he is more than 70 years old, but he said that he would spend the rest of his life on it.

Zhu Wenli pays homage to classical porcelain in his own way, as do all craftsmen, i.e. through assiduous study and aspiration for perfection.

Zhu Wenli has always molded the clay for ceramic wares of various shapes and sizes all by himself. Today, he is going to glaze a new batch of wares after the clay molding. This liquid, khaki in color, is a glaze solution made of a dozen minerals. Zhu Wenli stirs it evenly, and ladles it into a bottle. He shakes the bottle, and pours the glaze solution by rotating the bottle at an angle of 360 degrees. Then he holds the bottom of the bottle with one hand, dips the clay base into the glaze solution and takes it out quickly.

Zhu Wenli has developed a new glazing technique involving double dipping. In the first dipping, a certain humidity shall be maintained for the second, i.e., the ware is taken out when it is half dry. If the humidity isn't enough, the glazing quality will be affected, i.e., bubbles will appear on the surface of the ware during the second dip.

After the glazing, Zhu Wenli will give the clay base a "polish": fixing the glaze. He fixes the uneven glaze with a knife and covers the unglazed areas with a brush dipped in glaze solution.

China's ancient ceramic art has created countless pieces of classical works. The Ru porcelain of the Northern Song Dynasty is regarded as the peak of this craftsmanship. It is a pity that it only existed for 20 years and then disappeared. For more than 800 years, ceramic artisans have been making painstaking efforts to reproduce the Ru porcelain, Zhu Wenli being one of them.

Reproducing Ru porcelain is the ultimate challenge for ceramic artisans of later generations.

Zhu Wenli was making another attempt at reproducing Ru porcelain. He carefully placed the glazed wares into the kiln chamber. The wares were placed in three layers with accurately controlled spacing between each, for wares at different positions receive different amounts of heat, which will further bring different firing effects. Placed into the kilns are also several pieces of *Huozhao*.

Zhu Wenli halts the heating at intervals to check the temperature increase by taking out and observing a *Huozhao*. The process is repeated to observe the phased changes in his wares.

Huozhao, also called *Huobiao*, is a piece of glazed clay for measuring the temperature inside the kiln and the heating status of the wares. It is of a roughly

Porcelain wares being fired in a kiln

round shape with a small hole in the middle, so that it can be picked up with a hook during the firing process and taken out for observation. At each firing, Zhu Wenli usually puts 16 *Huozhao* in the observation hole. Then he closes the door of the kiln, and ignites the fire. Huge flames burn inside the kiln, and the firing process begins.

The firing process of Ru porcelain consists of 12 stages which bring different color changes to the wares. During the whole eight hours of firing, Zhu Wenli remains beside the kiln. He takes out the bricks closing the kiln opening regularly, and observes the *Huozhao* inside. The control of the whole process is based on his own experience and judgment.

Every once in a while, Zhu Wenli will pick up a *Huozhao* from the kiln with a thin rod and observe the color change of the glaze after it cools down in a few seconds. Unlike other porcelain, the pea-green color of Ru porcelain will change gradually in half an hour with the drop of temperature. When the 16 *Huozhao* are picked out and placed in a time order, the color change is displayed clearly. The ceramic wares taken out at 1,110 degrees and 1,070 degrees are all of a pea-green color. It is not until the last temperature of 1,050 degrees is reached that the color changes from pea-green to celeste-blue. That is to say, it is not until the last moment that an artisan can tell whether the ware has taken on the celeste-blue color of Ru porcelain.

The color change of Ru porcelain in the kiln, or "kiln change," is the most amazing among all porcelains. How could the craftsmen of the Song Dynasty manage this to create perfect celeste-blue? Zhu Wenli always wonders about this.

Zhu has discovered a secret in the course of his firing experience: Unlike other types of porcelain which undergo only one chemical change (kiln change), Ru porcelain undergoes two kiln changes, a fact which is not recorded in any historical materials.

After the firing, it takes about eight hours for the porcelain wares to cool completely before the kiln can be opened. When the kiln is opened and the wares are slowly pulled out, a crisp crackling sound is heard. This is caused by the cracking of the glazed surface of the wares, which is a natural phenomenon in porcelain. There are two reasons for the crackling. One is that the extension of the clay along a certain direction during molding affects the arrangement of its molecules. The second is the different expansion coefficients between the clay and the glaze. The glazed surface, with a higher shrinkage rate after cooling, endows Ru porcelain with unique characteristics in the form of fine and irregular cracking lines distributed in the glaze layer. The cracking lines of each piece of porcelain are unique. The cracking sound lasts about ten minutes. Zhu Wenli puts on his glasses, picks up a piece, and observes it carefully.

In a firing in 1987, Zhu Wenli discovered unexpectedly that several of his wares exhibited the celeste-blue color of Ru porcelain.

This celeste-blue color looks plain, but it is elegant and rich in elements. The color, which is something between blue and green, is soothing to both the eye and the mind, yet it is hard to describe in word. What's more, the color exhibits changes in different lighting and at different angles. In bright sunshine, the soft and smooth celeste color will exhibit a tender and glossy yellow tone. When observed under a magnifying glass, the glaze layer exhibits sparse bubbles, which resemble the stars scattered in the sky on an early autumn morning, being clear yet not cold, and exhibiting a charm without banality. Ru porcelain, with its pure, elegant and natural beauty, has become the unparalleled top porcelain of great value. It is the quintessence of clay, fire and superb craftsmanship.

Zhu Wenli examines a Ru porcelain ware.

Zhu said that only the color of the blue sky on horizon after rain can be called celeste-blue.

Zhu Wenli has brought a rare porcelain that had disappeared for 800 years back to the world. This is a miracle that has amazed the industry. Zhu is now a famous craftsman because of this achievement.

The story of Zhu Wenli started in the early 1980s, when the Ministry of Light Industry decided to implement the "study of celeste glaze in Ru porcelain" project. Technicians in the Ru porcelain factory, Zhu Wenli included, worked hard on the project. However, their enthusiasm and drive were baffled by the practical problems. The project was scheduled to be accomplished in half a year, but it still had not been completed even after two years. The project team was disbanded. Zhu Wenli was disappointed, but he decided to start all over again by himself. He was as obsessed with the project as if driven by some supernatural force. He studied day and night at home and at his own cost. Two years passed, and he did not succeed. He was so frustrated that he even considered giving up.

Just as he was in the depth of despair, he came across a myth in a book. It was said that Emperor Shizong (921–959) of the Later Zhou Dynasty ordered the building of the Chai kiln to produce celeste-blue porcelain. Two officers were

148

beheaded because of their failure. The third officer failed in his first two firings. If he failed in the third firing, he would be beheaded, too. When he returned home frustrated, his seven-year-old daughter asked, "Dad, why are you unhappy?" After being told the story, the little girl asked again: "How can we get the celeste-blue color?" The officer replied, "We must sacrifice a living person in the kiln. That is to say, somebody must jump into the kiln alive. But how can I sacrifice an innocent person just to make porcelain?" At the third firing, his daughter suddenly jumped into the kiln. The kiln burst open, and ceramic pieces flew out. These pieces were used for helmets and chest armor.

Zhu Wenli saw some hope in the myth, and decided to find out if there was any scientific basis in it. If anyone did jump into the kiln, there would be nothing left but bones. As the main constituents of bone are phosphorus and calcium, was it possible that these two elements played a key role in the firing process? Zhu then collected various types of bones and heated them using different methods. Nevertheless, all the burned bones were white, not celeste-blue color. Then he tried to add the burned bones to the glaze: 50 grams, 100 grams... He tried different amounts, but still failed to get the right color.

Although Zhu Wenli was deeply frustrated at the failures, he felt that there was always a mysterious force pushing him ahead whenever he thought of giving up. He drew lessons from his attempts at heating bones, and carried out more experiments, creating a total of 328 formulas. He tried, failed, and tried again. One day, he was firing another kiln following his latest formula. The process lasted two days and three nights. It was a small old kiln built in the 1980s, using an extremely high temperature, and Zhu had to drink a lot of water at the intervals of heating. Zhu stayed beside it all the time, guarding the kiln and heating it by himself. When he opened the kiln door, the high temperature of more than 1000 degrees blistered both his legs. He thought that he would succeed this time. However, when he observed a piece of *Huozhao* after the wares became vitrified, he was disappointed to see it was still a green color. Totally frustrated, Zhu Wenli went home and slept for several days. When he woke up, his first idea was to smash all the wares into pieces. When he opened the kiln, he discovered that almost all the wares were pea-green, but four of them had miraculously taken on the celeste-blue color. He was overcome with excitement.

Zhu Wenli smashes sub-standard wares.

In the following years, however, the celeste color that Zhu Wenli dreamed of never appeared in his kiln in large quantities. In nearly 40 years, he has made hundreds of thousands of ceramic wares, but there are only a few dozen refined ones which have taken on the perfect celeste color. He has produced various kinds of wares: lotus bowls, vessels in the shape of three cattle, thin-necked and round-shaped bottles, bottles with a long neck and two handles on each side, and so on. When grading the wares, Zhu uses a magnifying glass to observe them closely. If there is any flaw, such as uneven glaze or other slight deformation, it will not be deemed as top-class exquisite ware. There might not be a single top-class piece in an entire firing, and a qualification rate below 0.01% is common.

In the eyes of outsiders, Zhu Wenli, now in his seventies, is already a Ru porcelain master who truly deserves the name. However, Zhu Wenli thinks that it was merely luck that enabled him to make a few dozen celeste-blue wares in the past decades. He has not yet found out the real secret of Ru porcelain. In the more than 30 years since his first success, he has made many attempts but it seems that he has not made much progress.

Zhu Wenli molds clay.

A genuine craftsman will never deceive himself or seek fame by deceiving the world.

As for sub-standard porcelain wares, Zhu Wenli will smash them on the spot. He cannot remember how many times he has done that. Once he produced more than 200 wares in one month, but he smashed all of them. Each fragment carries Zhu Wenli's efforts. This is the price of perfection.

How can he uncover the ancient secret to ensure 100% success in reproducing Ru porcelain?

Every day Zhu Wenli carries a canvas bag and goes around Ruzhou city on his old bicycle. In the late 1970s, he bought his first bike, and has used 12 bikes in total in more than 30 years. The world has changed, but Zhu Wenli has not. Ruzhou city, or "porcelain city," was the center of production of the official porcelain wares of the Northern Song Dynasty. The Ru official kiln, the Northern Song Dynasty official kiln used to be located there. There are many ceramic items labeled "Ru porcelain" displayed in shops in the city, but in Zhu's eyes, these items, which use chemical pigments, are fakes. He also believes that even his most successful wares, which he has selected carefully, are inferior to the Ru porcelain of the Song Dynasty.

Zhu Wenli believes that the secret of Ru porcelain must be hidden underground, in ancient caves. Today, the ruins of the Northern Song Dynasty have long been buried. As Zhu Wenli cannot afford the excavation, he goes to various construction sites frequently. As long as porcelain pieces are unearthed on any site, he will arrive at the spot promptly, however severe the weather. Zhu often visits several construction sites in a single day, carrying a fork, shovel and brush. He digs carefully and puts each and every piece he finds, however small, into his bag.

Over the past 40 years, Zhu has frequented numerous construction sites, and discovered a dozen ancient kiln sites. However, he has not discovered the kiln which specialized in celeste-blue porcelain and which disappeared 800 years ago. He is no longer young, but his determination has never changed. Many people in Ruzhou know this "hardcore porcelain fan."

Zhu Wenli takes the ceramic pieces he has collected on construction sites to his studio. He observes carefully the subtle features of each piece under a magnifying glass. He also consults ancient books on the subject. He spends much time in the studio every day studying the porcelain. For any piece of porcelain, he can tell at a glance whether it is from a folk kiln or an official kiln, from which dynasty. Only porcelain has been able to keep his undivided research for decades. Zhu Wenli explained that the value of Ru porcelain lies in its glaze color, but technology is still unable to solve its mystery.

After years of exploration and comparative research, Zhu Wenli concluded that the major cause of his failure might lie in the ingredients of the glaze. But even if this conclusion is correct, it will not help, because the formula for Ru porcelain glaze has long been lost, together with the disappearance of the porcelain wares themselves. The glaze determines whether the color of Ru porcelain is authentic. As the glaze is composed of dozens of ores, it is difficult to figure out the components accurately, despite thousands of experiments.

In Zhu Wenli's glaze lab, there are a dozen bowls filled with various ore powders. When he prepares a glaze, he weighs the ingredients carefully. There must not be any mistake, for the slightest deviation means failure. Zhu has made thousands of experiments in the hope of finding the lost formula, but he has not succeeded yet. He is sure that the slight difference between his porcelain and that of the Song

Dynasty is caused by one ingredient. But what is it? Zhu has been searching hard. Every day he ponders the same question: How can he find the last ingredient?

In Zhu Wenli's opinion, as Ruzhou is the only area that produced Ru porcelain, due to the inconvenient transportation in ancient times, the artisans must have collected all the materials for the porcelain on the spot, i.e., near the kiln yard. Therefore, the ores used for Ru porcelain will only be found in the nearby mountains.

Ever since Zhu Wenli got this idea, he has been going into the mountains every three days on foot or by bike, depending the distance. With a canvas bag on his left shoulder and a small hammer in his right hand, he proceeds alone in the mountains. He has been to wheat fields, streams, mountains and forests to study the color and texture of ores.

Zhu Wenli has climbed all the mountains around Ruzhou. He usually sets out before dawn. At noon, he will find a place to rest and have his meager lunch of steamed buns and water, and go on with his work again. He doesn't go back until it is dark. He has been doing this for years, although he has not succeeded in finding what he is looking for yet.

Zhu Wenli fixes the clay bases.

Nowadays, Zhu Wenli's two daughters, who have inherited his quest, accompany him in his search. They always have a lot of discoveries. Can the ores they discover be used in the glaze? Will a miracle happen this time? Only the kilns can tell.

After a nonstop firing of three days and two nights, it was time to check the result. Zhu Wenli, exhausted, went up to check his wares as usual. He picked up the wares and observed them. However, the wares turned out to be failures again, so he smashed them one by one. It seemed that the wares were calling out, begging him to stop. Zhu Wenli felt that he was so close, yet so far away from success. An indescribable sense of urgency gripped the man in his 70s.

Again, he went to the official kiln of Ru porcelain that he discovered on a construction site in 1999. In his opinion, the discovery was the most fortunate moment in his life, as well as the closest to the great ancient artisans, so close, it seemed, that he could even touch them. The kiln was immediately protected by local people after the discovery. There were five holes on the kiln site, two big ones and three small ones, where pieces of ceramics were still gleaming faintly in the soil.

From then on, Zhu Wenli comes here before every important experiment and firing or whenever he is frustrated or faltering. He gets down to the bottom of the kiln and listens to any sound that seems to come from ancient times. He feels that there will come a day when the ancient craftsmen and he can hear each other. Zhu has been persisting and waiting all his life. In his constant pursuit and search, he dreams that a miracle will come when he opens his kiln someday.

This is Zhu Wenli's life. Throughout the ages, craftsmen may all have been those who persist in searching and are good at discovering. However long or difficult the search is, they will not be discouraged or give up their quest.

A Craftsman of Great Persistence

I first met Zhu Wenli on April 29, 2016, in Zhu's porcelain exhibition lounge. Zhu Wenli was plainly dressed. As I was not quite familiar with his Ruzhou dialect, I paid more attention to his expressions and body languages. In the three-hour interview, Zhu Wenli's hands were occupied by pieces of ceramics and books. He had a quite serious expression. It would not be exaggerating to describe his passion for Ru porcelain as "craziness" or "infatuation." Probably because he talked so much, I came to gradually understand his accent. One thing that impressed me most was that whatever the topics we started with, he would finally wind up with Ru porcelain. The old saying "One cannot say three sentences without mentioning his trade" would not suffice to describe Zhu Wenli, for everything he said was about Ru porcelain, as if there was nothing but porcelain in his life. When it came to the topic, his words were like surging water with no end.

Zhu Wenli is the top master of Ru porcelain. There is a saying that great wealth in the family is not worth a single piece of Ru porcelain. Zhu Wenli should be a very rich man, because he can sell his Ru porcelain for money. But in fact he isn't rich. He still works in a rented workshop, a plain old building built in the 1990s and sparsely decorated. His kiln is very shabby, too, but the wares he makes are better than those his elder daughter Zhu Yufeng makes in a new kiln, which further proves his superb skills. Zhu Yufeng's workshop is also rented, but it has better decoration. Zhu Yufeng asked her father to redecorate his workshop, but Zhu Wenli refused her every time. He said, "I live by my skills, not useless appearance."

In the courtyard of his workshop, there is a "porcelain tomb," where the pieces of porcelain he smashes are buried. Zhu Wenli is a genuine craftsman, but never a merchant. He told me that Ru porcelain was like his life. He can give up anything except Ru porcelain. He is obsessed with Ru porcelain above everything, and his goal is to bring perfect Ru porcelain back to the world. Therefore, he will smash any defective ceramic ware. In our filming of the program, we witnessed Zhu Wenli smashing a total of six kilns-worth of wares, or more than 200 pieces. There is a practice called "kiln wholesale" in Ruzhou, by which many factories sell a whole kiln-worth of wares they produce, whatever the quality. As for products of a quality similar to those of Zhu Wenli, one kiln-worth will be sold for at least 100,000 yuan. Zhu Wenli never engages in this kind of "kiln wholesale." He will not sell his best-quality products easily, unless he absolutely needs the money.

I was deeply impressed on the first day of filming, when Zhu Wenli went to the mountains to look for ores. All the shots on the mountain were to be completed in one day. In order to get better results, I told him in advance that we would climb four mountains that day. If he felt tired, we could have a rest half way. I was worried whether the 70-year-old was up to such strenuous physical activity. Our photographer said that even he could barely manage to climb four mountains in a day, let alone a 70-year-old. However, it turned out that Zhu Wenli was the fastest among the eight of us. He was busy climbing mountains, collecting ores and cooperating with us in the filming that day without taking any rest. In fact, he seemed to enjoy it very much.

Zhu Wenli's stamina comes from his decades of searching for ores. He wears cloth shoes every day, ready to climb mountains or go to construction sites at any time. Even at lunchtime, he eats quickly and then gets on with his study of porcelain. He never chats with others or wastes his time on things other than porcelain.

I was not familiar with Ru porcelain before, and didn't expect it to be so difficult to produce such a single-color porcelain. As I learnt more, I understood that defects in single-color porcelain were more obvious, even if it was only a bubble in the glaze layer. A ceramic ware with any obvious defect can never be a quality product, which explains why Zhu Wenli has produced only several dozen high-quality wares in so many years. In the firing process, an artisan recognizes the temperature change inside the kiln by observing the *Huozhao*. During the filming, I saw the last *Huozhao* Zhu Wenli picked out was of a pea-green color, but the color changed greatly in an instant. It was approaching blue, but still was not celeste-blue. I wondered if it could change into celeste-blue, and how long that would take. The next morning, about 12 hours after the kiln had been opened, the last *Huozhao* did indeed turn into the celeste-blue color. Only after I witnessed the entire process, did I believe the magical change.

Zhu Wenli has a very valuable quality in him. As a great craftsman who is recognized by Chinese and foreign experts and scholars, he constantly questions his own research, and denies his achievements again and again. He has never stopped in his pursuit, but spent his entire life exploring the footprints of his predecessors. He thought that the celeste-blue color in his porcelain was slightly different from that of the Song Dynasty. I made a careful comparison between the two, and felt that there was indeed a difference. The porcelain of the Song Dynasty, whatever the ware was, was thinner than today's wares. But the most important difference is the color. The glaze of Song porcelain is of a transparent blue color which I have never seen in any modern ware. It is the real celeste-blue color, a feast for the eyes and a joy for the heart.

Zhu Wenli said: "Once a writer paid me a special visit. He said that when he was tired of writing late at night, he would pick up a Ru porcelain ware and admire it against light, whereupon his mind would gradually become calm and lucid. Ru porcelain does have this function, and you can have a try. When we are appreciating the Ru porcelain, it is looking at us, too." At any time, in any place, in front of anyone, he will "advertise" Ru porcelain. Zhu Wenli, an old man who is obsessed with Ru porcelain, is making his porcelain with all his passion and life. Years spent with Ru porcelain has given him a heart of absolute persistence.

By YANG JING, documentary director

ZHANG DONGMEI

Chief Technician of Tongrentang

Profile

Zhang Dongmei, chief technician of Beijing Tongrentang Group, is a national model worker and inheritor of Tongrentang's "intangible cultural heritage" Angong Niuhuang bolus. Ever since entering the pharmaceutical factory of Tongrentang at the age of 17, Zhang Dongmei has been working there for more than 30 years, earnestly practicing the ancient medical motto of "Never saving labor in complicated processing, nor subtracting valuable ingredients from medicine." As the flagship product of Tongrentang, Angong Niuhuang bolus is a proprietary Chinese medicine with a history of more than 200 years. It is a "life-saving medicine" in the minds of many Chinese. Now, Tongrentang and Zhang Dongmei, the leader of the Angong Niuhuang Bolus Team with 26 members, have assumed the duty of inheriting this national intangible cultural heritage. At Tongrentang's production base in Beijing, Zhang and her team will show the special skill of making this medicine manually to thousands of visitors from about 80 countries and regions around the world every year.

Zhang Dongmei, chief technician of Beijing Tongrentang Group

Preview

More than 30 years ago, Zhang Dongmei succeeded her mother at Tongrentang. Three decades is a long time, yet it is a fleeting period when one looks back on it. Over the years, she has been engaging in the production of traditional Chinese medical boluses under the guidance of the ancient medical motto of "Never saving labor in complicated processing, nor subtracting valuable ingredients from medicine." Her work involves dosing, mixing, bolus formation, interior packaging, wax dipping, stamping and exterior packaging. It seems simple, but the acquisition of craft is not as simple as what is imagined.

Angong Niuhuang bolus, which weighs only three grams, is regarded as a life-saving "magic medicine." With a history of 213 years, this medicine is more valuable than gold. It is still made by hand in the traditional way, and is ranked the first of the "three treasures for curing fever-related diseases."

Besides precious ingredients, Angong Niuhuang bolus also features a complicated process: dosing, mixing, bolus formation, interior packaging, wax dipping, stamping and exterior packaging. The ready-made boluses must be of a round shape, smooth and lustrous in appearance, evenly colored and with a refined texture. The weight must be three grams per bolus, no more and no less.

Today we are going to meet the artisan who makes this magic bolus, Zhang Dongmei, the protagonist in this episode of *Great National Craftsmen*. She attains the above-mentioned rigid requirements completely through manual work, which is amazing.

"If you knew Zhang Dongmei's work style, you wouldn't be surprised at all," said Guo Fenghua, Zhang Dongmei's colleague for more than 20 years and deputy leader of the Angong Niuhuang Bolus Team.

Liang Bowei is Zhang Dongmei's apprentice. Three years ago, he came to work in Tongrentang after obtaining his master degree in the UK. As soon as he started work, he was deeply impressed by Zhang Dongmei's serious attitude and strict requirements. He felt that she was even more rigorous than his scientific research supervisor. Before entering the workshop, they must change their clothes twice and wash their hands three times. Zhang Dongmei breaks down hand washing into 18 actions: wet hands, wash fingers, check fingernails, wash wrists... Not a single step can be omitted. As the last step, one must apply alcohol to the hands for disinfection.

"Not only that, Ms. Zhang also asks other experienced colleagues to supervise us in case we miss any step," Liang Bowei said with a smile.

Why must one bother so much about merely washing hands?

"Angong Niuhuang bolus is made of 11 medicinal materials, including bezoar, cinnabar, musk and pearl. One of the primary standards for its production is 'pure material feeding.' All materials and production links must meet the hygiene standards, so as to ensure the quality and curative effect. For example, the animal hairs left in musk must be picked out one by one, without a single hair left. Experience and sense of responsibility are required in this regard. Angong

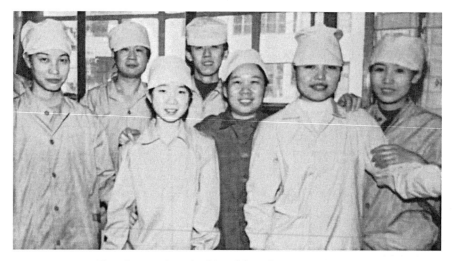

Zhang Dongmei (second right) with her colleagues in earlier years

Niuhuang bolus is the 'life-saving medicine' of our people, and there must not be the slightest deviation," Zhang Dongmei explained.

In order to guarantee the quality of the "life-saving medicine," Zhang exercises strict quality control. She pays attention to every operational detail, and has standardized all the operations in production. She insists that "product quality is ensured in production, but not tests," requiring the team to hold "producing 100% high quality products" as their mission and "going all-out for perfection" as the work criterion. She has launched the "three-level inspection" method of team sampling inspection, group patrolling inspection and individual self-inspection. For any problem discovered, everyone can exercise the responsibility of "quality inspector," ensuring that the quality of every production link of Angong Niuhuang bolus is under strict control. Ever since the production team was set up in 2004, the annual yield of one million boluses has had a 100% qualification rate.

Zhang Dongmei joined Tongrentang at the age of 17. Today, she is the chief technician of Tongrentang, and the only inheritor of the intangible heritage of Angong Niuhuang bolus. Now she maintains a 100% success rate in one-time manual bolus formation. However, she was not so good when she was an apprentice.

Zhang said, "At first, I could only make the raw material into a long strip, but couldn't make it into boluses. I failed no matter how much I tried."

162

"Strip-making" is the key step of manual bolus formation. It somewhat resembles making dough strips for dumplings. The strip should be of even thickness and appropriate length, or the speed and quality of bolus formation will be affected. For such a seemingly simple process, Zhang Dongmei practiced for more than a year.

Zhang recalled her past exeriences with a smile: "I was not very satisfied with the strips I made, so I made noodles when I got home, so that I could practice with the dough. In those days, my family ate noodles every day."

"The more difficult it was, the more I practiced." After Zhang acquired the skill of making strips, she had the opportunity to learn the critical step of making boluses.

"At first, I did not dare to make boluses, for fear that my bad performance might delay the work. Later, when the experts were working, I was asked to test the weight difference. I took the opportunity to observe how they worked. I practiced for many days, and finally learned the skill. My record is processing eight strips at the same time."

Bolus making requires strength. Zhang Dongmei basically makes the same gestures every day, with her spine straight and her arms raised. After a day's work, her back and waist often ache. Due to her intensive work, she has even suffered from a serious lumbar disc herniation. The first thing before she enters the workshop every day is to take medicine. "I take medicine subscribed by orthopedic doctors and painkillers. The doctors have no good solutions for this problem."

Apart from making boluses, gold foil wrapping, wax dipping, stamping and other links are all carried out manually. A few years ago, Tongrentang bought a bolus-making machine, but manual work is always indispensable. Zhang Dongmei said the boluses made manually were of high quality. It is impossible to rely on the machine completely, and the workers must have manual skills.

Angong Niuhuang bolus was invented by Wu Tang, a Qing Dynasty doctor specializing in treating febrile diseases. This medicine is effective for high-fever coma, stroke, cerebral hemorrhage and other severe and acute diseases. Angong Niuhuang bolus ranks the first among the top ten medicines of Tongrentang. The bolus contains eleven traditional Chinese medicines, including bezoar, ox horn powder, musk, pearl and coptis. The ingredients are ground into fine powder and mixed in certain proportions and then made into boluses. Tongrentang keeps this process highly confidential. Among the ingredients, the natural musk has

Zhang Dongmei talks with a customer at the Tongrentang pharmacy.

a resuscitation function. However, it used to be the biggest headache for Zhang Dongmei.

There are fine hairs in natural musk, which must be picked out by hand. It is a tedious and boring job that no machine can perform. There is no quality criterion in this hair-removing procedure, and it all depends on the worker's sense of duty as to how well the work is done.

Zhang said that a sense of duty was one of her requirements for her apprentices. Zhang Na, one of Zhang Dongmei's apprentices, has much to say in this regard: "When I picked out hairs for the first time, I repeated the process seven times. But my supervisor was not satisfied, and told me to do it again. I was quite depressed. Later, she told me that although no one would know if I didn't pick out all the hairs, we must do our utmost to produce the life-saving medicine. A maximum sense of responsibility has been a rule in the trade for generations."

Zhang Dongmei is always meticulous and striving for perfection in her job. She is strict with her apprentices, a virtue she has learned from her supervisor. Recalling the past, Zhang still remembers her supervisor's teachings: "When my supervisor and I were picking the hairs out of the musk, we each sat on a stool with

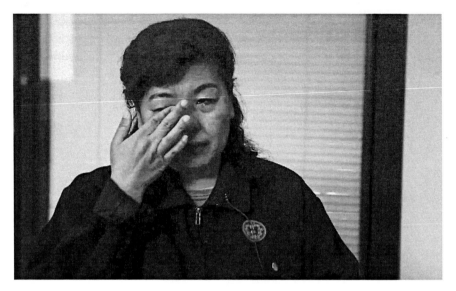

Zhang Dongmei sheds tears during an interview.

a big porcelain basin before us. After I had finished six or seven rounds, I asked if that was enough, but she said that it was far from being enough, and I had to keep going. I didn't expect it to be such strenuous work. My supervisor continued with my work after she had finished her own. It made me feel bad watching her doing my job bit by bit with great patience. Well, I changed my mind after that. I must make sure that I always complete my part of the work."

After this, Zhang Dongmei better understood Tongrentang's time-honored maxim of "Always work honestly in medicine making out of responsibility and conscience," as well as her own responsibility as a medicine maker.

Looking back on her career, Zhang said, "Throughout my life, I have been doing only one thing: making boluses. When my mother worked at Tongrentang, I often went to her work place with her when I was young. I told my mother that I would be happy to do my work well, because I didn't want to disappoint her."

At the Tongrentang pharmacy in Dashilan, Beijing, Zhang Dongmei receives customers from all over the country. A local customer who had come to buy Angong Niuhuang boluses said to her: "A friend in Nanjing asked me to buy the boluses for him. This is a life-saving medicine, and I thank you for your hard work."

When an elderly customer from Taiwan learned that Zhang Dongmei was the inheritor of the art of making Angong Niuhuang boluses, she bowed in admiration, and said to Zhang: "We must inherit and pass on traditional Chinese medicine to future generations. Thank you for what you have done!"

On hearing these words, Zhang Dongmei's eyes filled with tears. It was an affirmation of Chinese medicine as well as of herself. She felt that all her persistence had been worthwhile!

In Zhang's eyes, she is just doing what she should do. "Sometimes I wake up in the middle of the night, and try to remember if I have added the proper ingredients that day. I am satisfied if I can do my work well in terms of quality and quantity without making any mistakes. To put it simply, doing my work well during the day makes me sleep well at night."

Over the past 34 years, Zhang Dongmei has been doing only one job: making boluses in Tongrentang. However, during her years of persistence, she has been passing on the dedication and craftsmanship inherited in this craft over thousands of years.

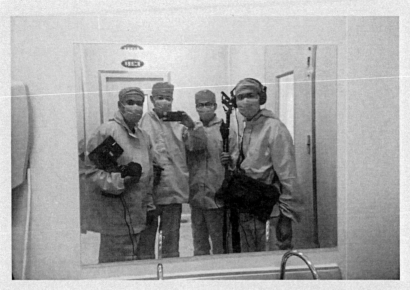

The "fully-armed" film crew at the workshop of Angong Niuhuang boluses

An Amiable and Optimistic Worker

It has been two years since the production of the episode of Zhang Dongmei as part of the series *Great National Craftsmen*. I always remember her as being amiable and optimistic.

Angong Niuhuang bolus ranks the first of the "three treasures for curing febrile diseases" of Tongrentang. The medicine is renowned for "treating emergent diseases with instant effect." In the eyes of many people, Zhang Dongmei included, Angong Niuhuang bolus is somewhat mysterious.

When we came to the Tongrentang pharmaceutical factory for the first time, Zhang Dongmei said, with a happy smile, that her apprentices all called her "Mom," probably because she liked chatting with them.

In our interview, entering the workshop was indeed a troublesome thing. As there were strict hygiene requirements, we had to wash our hands, put on sterile clothing and go through disinfection and other steps before entering. Take washing the hands for example. There were several sinks with different functions, and one had to wash one's hands in each of the basins in turn. Our cameras and other equipment went through several rounds of disinfection, too. As soon as we entered the workshop, Zhang Dongmei trotted ahead quickly and left us behind. Our cameraman had to remind her again and again to walk slowly. Zhang Dongmei said, "I'm sorry. I am used to walking quickly in the workshop..." She impressed us with her straightforwardness.

Zhang Dongmei succeeded her mother and began to work at Beijing Tongrentang Co., Ltd. in 1982. For more than 30 years Zhang has been working on the production line, engaging in the production of Chinese ready-made boluses. Her work includes dosing, mixing, bolus making (especially manually), etc. She has developed a keen sense of discernment and excellent skills. The production of Angong Niuhuang boluses is a very complicated process with extremely strict standards at each step.

Tongrentang has followed the traditional method in the production of Angong Niuhuang boluses. With Zhang and her colleagues' efforts, the yield is still considerable. The yield of the Yizhuang branch alone is 1 million boluses per year. Such a large yield, especially realized through a purely manual method, is a huge challenge for a team of only two dozen.

Zhang Dongmei, the head of the Angong Niuhuang bolus production team, said that in the past they had adopted the "flow line" production model, with staff specially assigned for each step: dosing, mixing, gold wrapping and packaging. Now, on the basis of a division of labor on the "flow line" model, a rotation system has been adopted. While ensuring the basic staff for each post, more staff are added to increase efficiency. For example, there used to be two workers for the gold wrapping process. Now under the rotation system, more workers are assigned to the procedure as appropriate to increase efficiency. When the gold wrapping process is completed, those workers can be assigned to the next working process. What used to be done by fewer workers is now done by more, and the output increased right from the start.

On the production line, a wooden mold that has been handed down for a hundred years is still indispensable in the bolus formation process. The mold, engraved with rows of pits, gives off a strong aroma of medicine due to years of contact with medicinal materials.

The workers put strips of medicinal material into the mold and put a small board on top of it. Then they press the board with both hands gently, and more than 20 small black boluses roll out of the mold. "The strength of the pressing is critical. One can only get the correct hand feel after a great deal of effort." Zhang Dongmei said. The ready-made boluses are weighed one by one. Each bolus should weigh 3 grams. According to national standards, the deviation shall be controlled within ± 0.21 of a gram. However, Tongrentang's standards are far stricter than the national standards. "We are really good at getting the exact weight!" said Zhang.

Then the workers wrap the boluses in gold foil. Gold, with its functions of calming and detoxication, is also an ingredient of Angong Niuhuang boluses, and must be wrapped around the boluses completely and evenly. There should not be a scrap of gold coming off, and the original shape of the bolus should not be changed. In the gold wrapping process, gold foil is issued to the workers every day, and no waste is allowed. The gold foil is so thin that it can be blown away by a breath, so workers must wear masks during this operation.

To keep them dry, the boluses wrapped in gold foil are wrapped again, this time in clear cellophane, and placed in two plastic hemispheres, to form plastic balls. Then the plastic balls are sealed with white wax and stamped with the characters signifying "Tongrentang Angong Niuhuang Bolus." This is the final step.

Throughout the entire production process, we only missed the procedure of hair picking, which Tongrentang keeps as a strict secret. In this process, workers pick out the fine hairs in the musk. This tedious work takes several hours, and can only be done manually. Zhang Dongmei has a deep impression of this procedure. She told us that she learned the skill from her supervisor when she was an apprentice. Once she spent several hours picking the hairs from musk, but was told that she was inexpert. Zhang Dongmei's narration of this story was the most touching part of the interview. As she recalled the story, she couldn't help shedding tears. The experience exerted a great influence on her. She said that the traditional Chinese medicine culture was passed down from generation to generation through such words and deeds. Ever since then, she had a deeper understanding of Tongrentang's time-honored maxim of "Always work honestly in medicine making out of responsibility and conscience" as well as her sacred responsibility as a bolus maker.

In the process of making boluses, Zhang Dongmei always had a serious expression. She told us that undivided attention should be maintained at each step. For example, when the raw material in the shape of long strips was to be pressed into boluses, the mold and the strips should be brushed with oil. The operation seemed simple, but it was quite demanding in that the operators had to ensure the weight of the boluses and meet the standards of them being "round, smooth and shiny" at the same time.

"As the raw materials for different medicines have different features, for example, viscosity, different approaches are adopted in the oil brushing," Zhang Dongmei said, with a serious expression. "We must learn by heart, and practice over and over again." She told us that many of the processes were highly demanding in the aspect of precision. Yet they could only be done manually, despite the risk of pinched fingers.

On November 17, 2015, Tongrentang officially established the "Zhang Dongmei Chief Technician Studio of Traditional Production Technology of Angong Niuhuang Bolus" and "Zhang Dongmei Innovation Studio." Henceforward, Zhang Dongmei was a formal "master medicine maker."

At a formal ceremony, Zhang Dongmei told her six apprentices that she was willing to teach them with great patience, and hoped that they could all stick to their posts and "work until retirement."

During the one-week interview, we worked with Zhang Dongmei every day. In those days, I deeply felt the persistence and responsibility of the workers of Tongrentang. Zhang Dongmei emphasized repeatedly that she "is only an ordinary person doing an ordinary job." As a national model worker and the inheritor of the intangible cultural heritage of Tongrentang Angong Niuhuang bolus, her greatest wish is to "make quality medicines for all patients."

By Lɪ Nɪɴɢ, reporter of
China Media Group & CCTV News Center

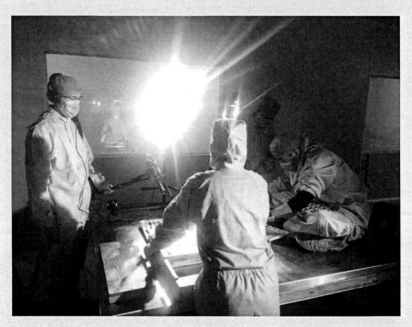

The videographer shoots the working scene of Zhang Dongmei.

BRIEF INTRODUCTION OF THE AUTHOR

The production team of *Great National Craftsmen* is composed by leading staff from different departments of the News Center of CCTV. They came to the workshops and production lines to talk with the workers and shoot their working scenes, showing the audience the real and vivid stories of the "Great National Craftsmen".